METHODS IN PHARMACOLOGY AND TOXICOLOGY

Series Editor
Y. James Kang
Department of Pharmacology and Toxicology
University of Louisville
Louisville, KY, USA

For further volumes:
http://www.springer.com/series/7653

Methods in Pharmacology and Toxicology publishes cutting-edge techniques, including methods, protocols, and other hands-on guidance and context, in all areas of pharmacological and toxicological research. Each book in the series offers time-tested laboratory protocols and expert navigation necessary to aid toxicologists and pharmaceutical scientists in laboratory testing and beyond. With an emphasis on details and practicality, *Methods in Pharmacology and Toxicology* focuses on topics with wide-ranging implications on human health in order to provide investigators with highly useful compendiums of key strategies and approaches to successful research in their respective areas of study and practice.

Immuno-Oncology

Cellular and Translational Approaches

Edited by

Seng-Lai Tan

Elstar Therapeutics, Cambridge, MA, USA

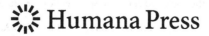

Editor
Seng-Lai Tan
Elstar Therapeutics
Cambridge, MA, USA

ISSN 1557-2153   ISSN 1940-6053 (electronic)
Methods in Pharmacology and Toxicology
ISBN 978-1-0716-0173-0   ISBN 978-1-0716-0171-6 (eBook)
https://doi.org/10.1007/978-1-0716-0171-6

© Springer Science+Business Media, LLC, part of Springer Nature 2020, corrected publication 2020
Chapter 9 is licensed under the terms of the Creative Commons Attribution 4.0 International License (http://creativecommons.org/licenses/by/4.0/). For further details see license information in the chapters.
This work is subject to copyright. All rights are reserved by the Publisher, whether the whole or part of the material is concerned, specifically the rights of translation, reprinting, reuse of illustrations, recitation, broadcasting, reproduction on microfilms or in any other physical way, and transmission or information storage and retrieval, electronic adaptation, computer software, or by similar or dissimilar methodology now known or hereafter developed.
The use of general descriptive names, registered names, trademarks, service marks, etc. in this publication does not imply, even in the absence of a specific statement, that such names are exempt from the relevant protective laws and regulations and therefore free for general use.
The publisher, the authors, and the editors are safe to assume that the advice and information in this book are believed to be true and accurate at the date of publication. Neither the publisher nor the authors or the editors give a warranty, express or implied, with respect to the material contained herein or for any errors or omissions that may have been made. The publisher remains neutral with regard to jurisdictional claims in published maps and institutional affiliations.

This Humana imprint is published by the registered company Springer Science+Business Media, LLC, part of Springer Nature.
The registered company address is: 233 Spring Street, New York, NY 10013, U.S.A.

Preface

Cancer would be quite common in long-lived organisms if not for the protective effects of immunity

Ehrlich P, Ned Tijdschr Geneeskd 5, 273 (1909)

The first wave of cancer immunotherapeutics, including cytokines, targeted and immune-modulating antibodies, and adoptive cellular immunotherapy, has resulted in long-lasting overall survival and potentially permanent protection against cancer recurrence. Unfortunately, only a small number of cancer patients respond favorably to current immune-based modalities, with some experiencing severe immune-related adverse events. Notwithstanding, the remarkable clinical success of immuno-oncology has sparked a new hope in our fight against cancer, with exponentially growing biotechnology and pharmaceutical efforts to search for the next generation of safer and more effective cancer immunotherapeutics aimed at broadening and prolonging patient response rates.

The main goal of this book is to serve as a general reference guide for identifying and applying cell-based translational assays commonly used as well as for assessing the therapeutic potential of new immuno-oncology therapeutics and advancing their mechanism of action. Given the high rates of clinical failures in oncology and the questionable predictive value of widely used animal models, increasing emphasis is being placed on the utility of in vitro and ex vivo translational assays using primary human cells. One noted paradigm shift away from the strict reliance on traditional cell-based assays and animal models in the biomedical industries is the application of primary cells, including human patient biosamples, that preserve or recapitulate as much as possible the essential human immune-tumor components observed in cancer patients.

It is also hoped that this book will provide readers with a baseline understanding of the pros and cons as well as key considerations for applying these assays that are more reflective of the human immune-tumor microenvironment to increase their translatability into the clinic. It is impossible for any of these assay systems to accurately model the various clinical tumor types and their relevance in terms of the immune contexture observed in cancer patients. Furthermore, patients enrolling in clinical trials usually have established cancers that are refractory to several previous lines of therapies that may result in compromised immunity and/or emergence of other immune escape mechanisms. Thus, as we improve our understanding of drug resistance mechanisms observed in patients, it is also envisioned that this book will serve as a starting point for further improvement and refinement of human translational assays that represent these scenarios for combinatorial drug screens and for developing precision medicines.

Cambridge, MA, USA *Seng-Lai Tan*

The original version of this book was revised. The correction to this book is available at https://doi.org/10.1007/978-1-0716-0171-6_14

Contents

Preface .... *v*
Contributors .... *ix*

1 An In Vitro System Combining Tumor Cells and Lymphocytes
  to Predict the In Vivo Response of Immunomodulatory Antibodies 1
  Min Dai, Ingegerd Hellstrom, and Karl Erik Hellstrom

2 High-Throughput Direct Cell Counting Method for
  Immuno-Oncology Functional Assays Using Image Cytometry 13
  Leo Li-Ying Chan

3 In Vitro Functional Assay Using Real-Time Cell Analysis
  for Assessing Cancer Immunotherapeutic Agents 35
  Biao Xi, Peifang Ye, Vita Golubovskaya, and Yama Abassi

4 Assessing the Potency of T Cell-Redirecting Therapeutics
  Using In Vitro Cancer Cell Killing Assays 51
  *Tomasz Dobrzycki, Andreea Ciuntu, Andrea Stacey, Joseph D. Dukes,
  and Andrew D. Whale*

5 Induction and Potential Reversal of a T Cell Exhaustion-Like State:
  In Vitro Potency Assay for Functional Screening of Immune Checkpoint
  Drug Candidates 73
  *Eden Kleiman, Wushouer Ouerkaxi, Marc Delcommenne,
  Geoffrey W. Stone, Paolo Serafini, Mayra Cruz Tleugabulova,
  and Pirouz M. Daftarian*

6 A Human In Vitro T Cell Exhaustion Model for Assessing
  Immuno-Oncology Therapies 89
  *Lynne S. Dunsford, Rosie H. Thoirs, Emma Rathbone,
  and Agapitos Patakas*

7 Validation of an Image-Based 3D Natural Killer Cell-Mediated
  Cytotoxicity Assay 103
  Brad Larson, Lubna Hussain, and Jenny Schroeder

8 3D-3-Culture: Tumor Models to Study Heterotypic Interactions
  in the Tumor Microenvironment 117
  *Sofia P. Rebelo, Catarina Pinto, Nuno Lopes, Tatiana R. Martins,
  Paula Marques Alves, and Catarina Brito*

9 Considerations in Developing Reporter Gene Bioassays for Biologics 131
  *Jamison Grailer, Richard A. Moravec, Zhijie Jey Cheng,
  Manuela Grassi, Vanessa Ott, Frank Fan, and Mei Cong*

10 Miniaturized Single Cell Imaging for Developing Immuno-Oncology
   Combinational Therapies 157
   Thomas Jacob, Pavani Malla, and Tania Vu

11 Precision Medicine: The Function of Receptor Occupancy in Drug Development......... 167
Leanne Flye-Blakemore, Christèle Gonneau, Nithianandan Selliah, Ajay Grover, Sriram Ramanan, Alan Lackey, and Yoav Peretz

12 In Vitro Assays for Assessing Potential Adverse Effects of Cancer Immunotherapeutics......... 199
Jinze Li, Mayur S. Mitra, and Gautham K. Rao

13 The Quest for the Next-Generation of Tumor Targets: Discovery and Prioritization in the Genomics Era......... 239
Leonardo Mirandola, Franco Marincola, Gianluca Rotino, Jose A. Figueroa, Fabio Grizzi, Robert Bresalier, and Maurizio Chiriva-Internati

Correction to: Considerations in Developing Reporter Gene Bioassays for Biologics......... C1

Index......... 255

Contributors

YAMA ABASSI • *ACEA Biosciences, A part of Agilent, San Diego, CA, USA*
PAULA MARQUES ALVES • *iBET, Instituto de Biologia Experimental e Tecnológica, Oeiras, Portugal; Instituto de Tecnologia Química e Biológica António Xavier, Oeiras, Portugal*
ROBERT BRESALIER • *Department of Gastroenterology, Hepatology and Nutrition, University of Texas MD Anderson Cancer Center, Houston, TX, USA*
CATARINA BRITO • *iBET, Instituto de Biologia Experimental e Tecnológica, Oeiras, Portugal; Instituto de Tecnologia Química e Biológica António Xavier, Oeiras, Portugal*
LEO LI-YING CHAN • *Nexcelom Bioscience LLC, Lawrence, MA, USA*
ZHIJIE JEY CHENG • *Promega R&D, Madison, WI, USA*
MAURIZIO CHIRIVA-INTERNATI • *Kiromic BioPharma, Houston, TX, USA; Department of Gastroenterology, Hepatology and Nutrition, University of Texas MD Anderson Cancer Center, Houston, TX, USA*
ANDREEA CIUNTU • *Immunocore Ltd., Oxfordshire, UK*
MEI CONG • *Promega R&D, Madison, WI, USA*
PIROUZ M. DAFTARIAN • *JSR Life Sciences LLC, Sunnyvale, CA, USA*
MIN DAI • *Department of Pathology, Harborview Medical Center, University of Washington, Seattle, WA, USA*
MARC DELCOMMENNE • *MBL International, Woburn, MA, USA*
TOMASZ DOBRZYCKI • *Immunocore Ltd., Oxfordshire, UK*
JOSEPH D. DUKES • *Immunocore Ltd., Oxfordshire, UK*
LYNNE S. DUNSFORD • *Antibody Analytics, BioCity Glasgow, Motherwell, Scotland, UK*
FRANK FAN • *Promega R&D, Madison, WI, USA*
JOSE A. FIGUEROA • *Kiromic BioPharma, Houston, TX, USA*
LEANNE FLYE-BLAKEMORE • *Covance Central Validation Services, Brentwood, TN, USA*
VITA GOLUBOVSKAYA • *ProMab Biotechnologies, Inc., Richmond, CA, USA*
CHRISTÈLE GONNEAU • *Covance Central Laboratory/Validation Services, Lausanne, Switzerland*
JAMISON GRAILER • *Promega R&D, Madison, WI, USA*
MANUELA GRASSI • *Promega R&D, Madison, WI, USA*
FABIO GRIZZI • *Department of Immunology and Inflammation, Humanitas Clinical and Research Center, Milan, Italy*
AJAY GROVER • *Flow Cytometry, Covance Central Laboratory/Validation Services, Indianapolis, IN, USA*
INGEGERD HELLSTROM • *Department of Pathology, Harborview Medical Center, University of Washington, Seattle, WA, USA*
KARL ERIK HELLSTROM • *Department of Pathology, Harborview Medical Center, University of Washington, Seattle, WA, USA*
LUBNA HUSSAIN • *Lonza Walkersville, Inc., Walkersville, MD, USA*
THOMAS JACOB • *Department of Biomedical Engineering, Knight Cancer Research Institute, Oregon Health and Science University, Portland, OR, USA*
EDEN KLEIMAN • *JSR Life Sciences LLC, Sunnyvale, CA, USA*
ALAN LACKEY • *Covance Central Validation Services, Brentwood, TN, USA*
BRAD LARSON • *BioTek Instruments, Inc., Winooski, VT, USA*

JINZE LI • *Department of Safety Assessment, Genentech Inc., South San Francisco, CA, USA*
NUNO LOPES • *iBET, Instituto de Biologia Experimental e Tecnológica, Oeiras, Portugal; Instituto de Tecnologia Química e Biológica António Xavier, Oeiras, Portugal*
PAVANI MALLA • *Department of Biomedical Engineering, Knight Cancer Research Institute, Oregon Health and Science University, Portland, OR, USA*
FRANCO MARINCOLA • *Kiromic BioPharma, Houston, TX, USA*
TATIANA R. MARTINS • *iBET, Instituto de Biologia Experimental e Tecnológica, Oeiras, Portugal; Instituto de Tecnologia Química e Biológica António Xavier, Oeiras, Portugal*
LEONARDO MIRANDOLA • *Kiromic BioPharma, Houston, TX, USA*
MAYUR S. MITRA • *Department of Safety Assessment, Genentech Inc., South San Francisco, CA, USA*
RICHARD A. MORAVEC • *Promega R&D, Madison, WI, USA*
VANESSA OTT • *Promega R&D, Madison, WI, USA*
WUSHOUER OUERKAXI • *MBL International, Woburn, MA, USA*
AGAPITOS PATAKAS • *Antibody Analytics, BioCity Glasgow, Motherwell, Scotland, UK*
YOAV PERETZ • *Covance Central Validation Services, Brentwood, TN, USA; Covance Central Laboratory Services, Indianapolis, IN, USA*
CATARINA PINTO • *iBET, Instituto de Biologia Experimental e Tecnológica, Oeiras, Portugal; Instituto de Tecnologia Química e Biológica António Xavier, Oeiras, Portugal*
SRIRAM RAMANAN • *Flow Cytometry, Covance Central Laboratory/Validation Services, Indianapolis, IN, USA*
GAUTHAM K. RAO • *Department of Safety Assessment, Genentech Inc., South San Francisco, CA, USA*
EMMA RATHBONE • *Antibody Analytics, BioCity Glasgow, Motherwell, Scotland, UK*
SOFIA P. REBELO • *iBET, Instituto de Biologia Experimental e Tecnológica, Oeiras, Portugal*
GIANLUCA ROTINO • *Kiromic BioPharma, Houston, TX, USA*
JENNY SCHROEDER • *Lonza Cologne GmbH, Cologne, Germany*
NITHIANANDAN SELLIAH • *Flow Cytometry, Covance Central Laboratory/Validation Services, Indianapolis, IN, USA*
PAOLO SERAFINI • *Department of Microbiology and Immunology, Miller School of Medicine, The University of Miami, Miami, FL, USA*
ANDREA STACEY • *Immunocore Ltd., Oxfordshire, UK*
GEOFFREY W. STONE • *NGM Biopharmaceuticals, South San Francisco, CA, USA*
ROSIE H. THOIRS • *Antibody Analytics, BioCity Glasgow, Motherwell, Scotland, UK*
MAYRA CRUZ TLEUGABULOVA • *JSR Life Sciences LLC, Sunnyvale, CA, USA*
TANIA VU • *Department of Biomedical Engineering, Knight Cancer Research Institute, Oregon Health and Science University, Portland, OR, USA*
ANDREW D. WHALE • *Immunocore Ltd., Oxfordshire, UK*
BIAO XI • *ACEA Biosciences, A part of Agilent, San Diego, CA, USA*
PEIFANG YE • *ACEA Biosciences, A part of Agilent, San Diego, CA, USA*

Chapter 1

An In Vitro System Combining Tumor Cells and Lymphocytes to Predict the In Vivo Response of Immunomodulatory Antibodies

Min Dai, Ingegerd Hellstrom, and Karl Erik Hellstrom

Abstract

We have developed an in vitro system to predict the in vivo response when treating cancer by administering monoclonal antibodies (mAbs) to checkpoint inhibitors (Dai et al., J Immunother 39(8):298–305, 2016). In this article we review our major findings could fulfill a clinical need.

Key words Checkpoint inhibitors, Immunotherapeutic agents, Monoclonal antibodies (mAbs), Peripheral blood mononuclear cells (PBMCs), Th1 response, Th2 response, Mouse tumor models, mAb combination

1 Background

Immunological approaches are revolutionizing cancer therapy [1]. So far, the most impressive effects have been with mAbs to checkpoint inhibitors which have produced complete tumor remission both in mice [2–6] and humans [7–15]. Complete remissions and potential cures have been obtained, particularly with melanoma but also with some other tumors including non-small cell lung carcinoma. However, most patients are not cured and it is often hard to foresee who will respond and side-effects are common [8, 16–18]. An in vitro model that can predict the response would be helpful and may be useful also to explore other immunotherapeutic agents.

Our group has cultivated lymphoid cells from tumor-bearing mice and cancer patients together with syngeneic (for mice) or autologous (for humans) tumor cells in the presence of immunomodulatory mAbs to generate tumor cell destructive immune responses [19]. As observed in tumor-bearing mice [5, 6, 20], a tumoricidal response was generated and was associated with a shift to a Th1 type immunity with a dramatic increase of CD3, CD4, and

CD8 T cells, including long-term memory T cells and lymphocytes expressing TNFγ and there was a decrease of CD19+ cells, myeloid-derived suppressor cells, and regulatory T cells [19]. Analogous results were obtained in pilot experiments with peripheral blood leukocytes (PBMC) from patients with ovarian carcinoma.

2 Experimental Procedures for Studies in Mice

C57BL/6 females (from the Jackson Laboratory) were transplanted subcutaneously (s.c.) with 5×10^5 cells from cultured TC1 mouse lung carcinoma, 1×10^5 cells from the B16 melanoma or intraperitoneally with 3×10^6 cells from the ID8 ovarian carcinoma [6], and C3H females (from the Jackson Laboratory) were transplanted s.c. with 5×10^5 cells from the SW1 clone of the K1735 melanoma. All murine tumor cell lines were obtained from ATCC. Five days later, spleen cells from these mice were cultured together with cells from the respective tumors in the presence of various combinations of immunomodulatory mAbs. The in vitro data were compared with survival data from mice where the same mAb combinations had been injected intratumorally. Antibodies to CD137 (LOB12.3; Cat. #BE0169), CTLA4 (9D9; Cat. #BE0164), PD-1(RMP1-14; Cat. #BE0146), and CD19 (1D3; Cat. #BE0150) were bought from BioXcell (West Lebanon, NH) and conjugated to magnetic beads as previously described [21, 22]. Beads that had been conjugated with an irrelevant mAb (2A3; Cat. #BE0089) were used as control and administered so that there was approximately the same amount of mAb protein in experimental and control groups. For conjugation, 1×10^8 Dynabead® M-450 beads were incubated with 50 μg of each mAb for 16 h, after which they were washed once with sterile PBS/BSA and suspended at 1×10^8 beads/ml.

Cultured tumor cells were seeded at a density of 5×10^4 cells/well into 24-well tissue culture plates (Corning Inc., Corning, NY) which contained 2 ml RPMI medium (Gibco, Rockville, MD) and 10% fetal calf serum (Atlanta Biological, Norcross, GA). Following overnight incubation at 37 °C in a 5% CO_2 in air atmosphere, 1×10^6 lymphoid cells were added to each well together with 3×10^6 mAb conjugated beads for each mAb per well. Following 5 days of cocultivation, the lymphoid cells and beads were removed from the culture wells, photographs were taken of wells in experimental groups and controls, and the number of tumor cells per well was determined by the MTS assay (Progema, Madison, WI). For the MTS assay, we pipetted 100 μl of the combined MTS/PMS solution into each well which contained the cultured cells in 500 μl medium and recorded the absorbance at 490 nm using an ELISA plate reader after 1 h incubation. The number of tumor cells/well was calculated from the OD values, and percentage of killed tumor cells was determined by comparing with the number of tumor cells in the control.

For quantitative PCR the total RNA was extracted from cocultivated lymphocytes using Qiagen RNeasy Mini Kit, followed by cDNA synthesis using iScript™ Reverse Transcription Supermix (Bio-rad, Hercules, CA). Subsequently, cDNA was used to measure the mRNA level of IFNγ and Tbx21 using qPCR on ABI Prism 7900 Sequence Detector System (Applied Biosystems). The relative quantification was performed using the comparative CT method as described by the manufacturer.

For statistical analysis, results were expressed as mean ± SEM. Student's test was used to compare the statistical difference between two groups, and one-way ANOVA was used to compare three or more groups. A correlation analysis was performed using GraphPad Prism 5.

3 Experimental Procedures for Studies with Human Cancer

Human serous ovarian carcinoma cells from two cell lines, 272 and 225, which had been established in our laboratory, were seeded, 5×10^4 cells/well, into 24-well tissue culture plates (Corning Inc., Corning, NY) where each well contained 2 ml RPMI medium (Gibco, Rockville, MD) and 10% fetal calf serum (Atlanta Biological, Norcross, GA). Following overnight incubation at 37 °C in a 5% CO_2 in air atmosphere, frozen and stored peripheral blood mononuclear cells (PBMC) from the autologous patient were thawed. Subsequently, dead cells were removed using a Dead Cell Removal Kit (Miltenyi Biotec) and 1×10^6 PBMC were added per well together with 3×10^6 beads which had been conjugated with various immunomodulatory mAbs using a mAb to an irrelevant antigen as control. As in the experiments with mouse tumors, the mixture was cocultivated for 5 days after which the lymphoid cells and beads were removed, photographs taken, and the number of tumor cells per well determined by the MTS assay (Progema, Madison, WI).

For flow cytometry, isolated lymphoid cells were washed with FACS staining buffer and incubated with human Fc receptor binding inhibitor for 10 min before staining for 30 min with mAbs to CD3 (clone OKT3), CD19 (clone HIB19), CD8 (clone RPA-T8), CD4 (clone RPA-T4), Foxp3 (clone PCH101), CD11c (clone 3.9), and CD83 (clone HB-15e); all bought from eBioscience, San Diego, CA. We used a FACSCalibur (BD Biosciences) machine and the data were analyzed using Flow Jo software (Tree Star, Ashland, OR). qPCR and statistics analyses were performed as described for the mouse experiments.

4 Major Findings with Mouse Tumors

As shown in Fig. 1, the majority of TC1 lung carcinoma cells was destroyed after cultivation for 5 days together with splenocytes from a mouse with a small TC1 tumor and magnetic beads conjugated with anti-CD137/PD-1/CTLA4/CD19 mAbs (referred to as "the 4 mAb combination"). The number of spleen lymphocytes

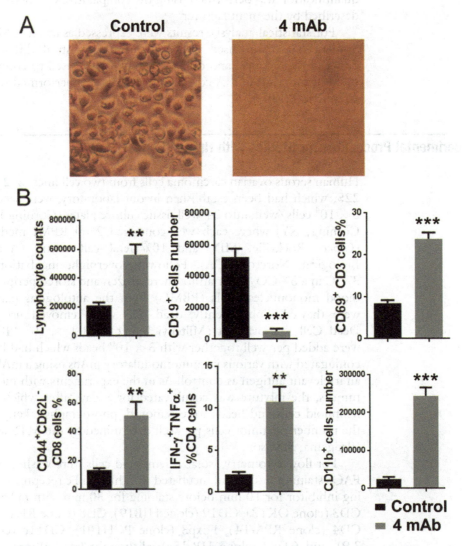

Fig. 1 (**a**) The photograph shows killing of TC1 mouse lung carcinoma cells by spleen cells from a mouse with a small such tumor which had been cultivated for 5 days with TC1 cells in the presence of beads which had been conjugated with the 4 mAb combination (anti-CD137/PD-1/CTLA4/CD19). (**b**) The 4 mAb combination stimulated proliferation of lymphocytes, decreased the number of CD19$^+$ cells, increased the number of CD3$^+$ cells expressing the CD69 activation marker as well as of CD8$^+$ long-term memory T cells, IFNγ$^+$ TNFα$^+$ CD4$^+$ cells, and cells expressing the CD11b$^+$ activation marker. $n = 5$ for each group, $^*p < 0.05$, $^{**}p < 0.01$, $^{****}p < 0.0001$

from these cultures was approximately 2.5 times higher than in the control where the beads were conjugated with an irrelevant mAb. The number of CD19$^+$ cells dramatically decreased while the number of T cells expressing the activation markers CD11b or CD69 increased as did CD44$^+$CD62L$^-$ long-term memory CD8$^+$ T cells and CD4$^+$ T cells expressing IFNγ and TNFα. The phenotype of the in vitro stimulated spleen cells was thus of the Th1 type as were spleen cells from mice which rejected established TC1 tumors after treatment with the 4 mAb combination [6, 20]. Cells from tumor draining lymph nodes likewise generated a tumor cell destructive Th1 response when cultivated with TC1 cells in the presence of the 4 mAb combination [19].

Figure 2 compares the ability of mAb combinations to generate a tumor destructive immune response in vitro with the response in vivo to intra-tumor injection of combinations of immunomodulatory mAbs. Figure 2a shows the mean survival time of treated

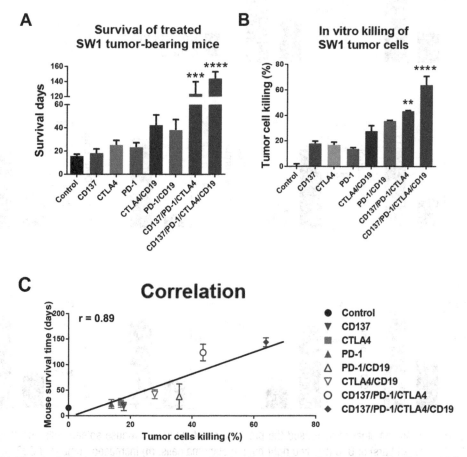

Fig. 2 (**a**) Survival of mice with SW1 melanoma after treatment with various mAb combinations which started when the tumors had a diameter of approximately 5 mm. (**b**) In vitro killing of SW1 cells by spleen cells from a mouse with a small SW1 tumor in the presence of the indicated mAb combinations. (**c**) Correlation between the in vivo and in vitro findings with various combinations of immunomodulatory mAbs. $n = 5$ for each group, *$p < 0.05$, **$p < 0.01$, ***$p < 0.001$, ****$p < 0.0001$

mice with SW1 melanoma, Fig. 2b depicts the numbers of surviving SW1 cells after coculture with splenocytes and beads conjugated with the respective mAbs and Fig. 2c shows that there was a statistically significant correlation between the in vivo and in vitro data. A similar correlation was obtained with the TC1 lung carcinoma and the B16 melanoma, i.e., the in vitro sensitization assay significantly predicted the therapeutic activity. The 4 mAb combination (anti-CD137/PD-1/CTLA4/CD19) was most therapeutically efficacious. As shown in Fig. 3a, it increased the number of lymphocytes that had been cocultured with spleen cells from a mouse with SW1 (left part) or B16 (right part) tumor. Figure 3b

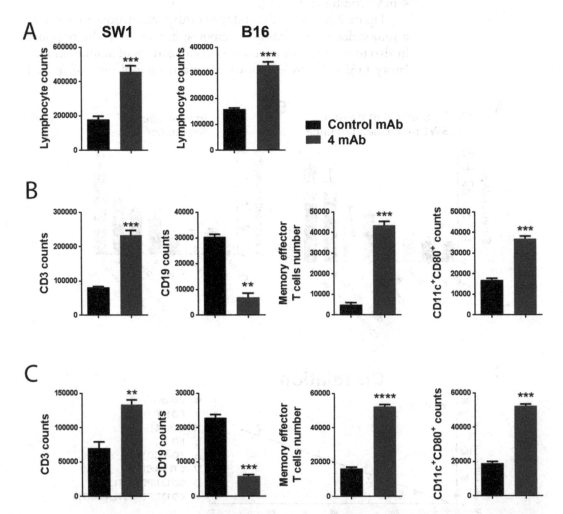

Fig. 3 (**a**) The 4 mAb combination increased the proliferation of syngeneic mouse spleen cells cocultivated with SW1 (the two left bars) or B16 (the two right bars) melanoma cells. (**b**) Increased number of CD3$^+$ cells, CD44$^+$CD62L$^-$ memory T cells, and CD11C$^+$CD80$^+$ mature dendritic cells and decreased number of CD19 cells in experiments with the SW1 melanoma. (**c**) Increased number of CD3$^+$ cells, CD44$^+$CD62L$^-$ memory T cells, and CD11C$^+$CD80$^+$ mature dendritic cells and decreased number of CD19$^+$ cells in experiments with the B16 melanoma. $n = 5$ for each group, $^*p < 0.05$, $^{**}p < 0.01$, $^{***}p < 0.001$, $^{****}p < 0.0001$

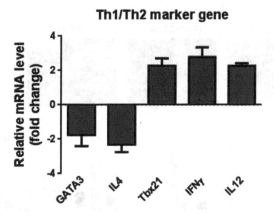

Fig. 4 Increased expression of Th1 related genes and decreased expression of Th2 related genes in mouse spleen cells from TC1 tumor-bearers cocultivated with TC1 cells in the presence of the 4 mAb combination

shows data from an experiment with the SW1 melanoma, and similar data with the B16 melanoma are shown in Fig. 3c. For both tumors there was an increase of $CD3^+$ cells, memory T cells, and mature dendritic cells and a decrease of $CD19^+$ cells. These findings are similar to those obtained with spleens and tumor-draining lymph nodes from mice that have been successfully treated in vivo with the 4 mAb combination [6, 20].

The PCR experiment in Fig. 4 demonstrates that the 4 mAb combination increased the expression of the Tbx21, IFNγ and IL12 genes which are characteristic of a Th1 type response [23, 24] and decreased expression of the GATA3 and IL4 genes which are characteristic of a Th2 response. The data are thus similar to those from mice responding to in vivo administration of immunomodulatory mAbs [5, 6].

5 Major Findings from Studies with Human Material

As shown in the photos and bar graph in Fig. 5a, most cells from the 225 human ovarian carcinoma were destroyed after cocultivation for 5 days with autologous PBMC in the presence of beads conjugated with the 4 mAb combination, and approximately 50% of the cells were killed when anti-CTLA4 plus anti-PD1 was used for checkpoint inhibition. Figure 5 panel (b) shows that, as in the mouse experiments, there was an increased expression of Th1 related gene (Tbx21, IFNγ, and IL12) and decreased expression of Th2 genes (Gata3 and IL4) when the PBMC were cultivated with autologous tumor cells in the presence of the 4 mAb combination, while the effect of the anti-CTLA4/PD1 combination was modest. The numbers of $CD3^+$, $CD4^+$, $CD8^+$, and $CD11C^+CD83^+$ mature dendritic cells increased, and the number

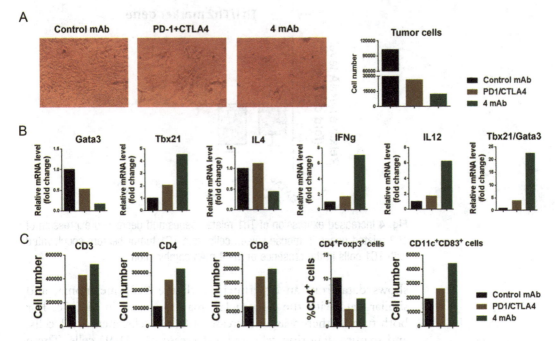

Fig. 5 (**a**) Killing of human ovarian carcinoma 225 cells cocultivated with autologous PBMC in the presence of anti-PD1/CTLA4 mAbs or the 4 mAb combination which is even more effective. (**b**) Increased expression of Tbx2, IFNγ and IL12 genes and decreased expression of Gata3 and IL4 genes by PBMCs which had been cocultivated with autologous 225 cells in the presence of the 4 mAb combination while the immunological effects with anti-PD-1/CTLA4 are less prominent. (**c**) Increased numbers of CD3$^+$, CD4$^+$, CD8$^+$, and CD11C$^+$CD83$^+$ cells and decreased numbers of CD4$^+$Foxp3$^+$ Tregs after cocultivation of 225 ovarian carcinoma cells with autologous PBMC in the presence of either anti-PD-1/CTLA4 mAbs or the 4 mAb combination

of CD4$^+$Foxp3$^+$ Tregs decreased with the 4 mAb combination being most effective. The phenotype of the in vitro stimulated PBMCs was thus of the Th1 type.

The results were similar for the 272 ovarian carcinoma, as shown in Fig. 6 where the tumor cells and PBMC were cocultivated in the presence of the 4 mAb combination (or control).

6 Future Studies

In a recent experiment, spleen cells were sensitized from a naïve mouse, from a mouse with a small tumor (approximately 3 mm diameter) and from one with a tumor of 8–10 mm diameter. As shown in Fig. 7, the tumor cell killing was much stronger with spleen cells from the mouse with the large tumor and this was reflected by a higher expression of genes encoding IFNγ and IL12 in response to stimulation with each of three different immunomodulating mAbs. Major effort will be given to extend these experiments to learn more about the influence of tumor size,

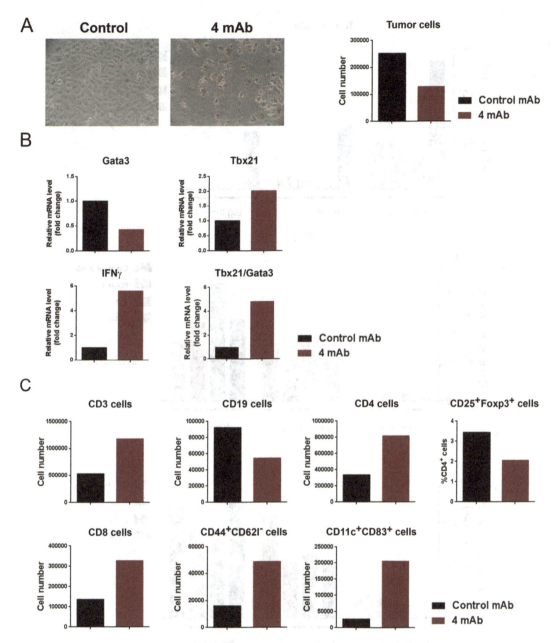

Fig. 6 (a) Killing of cells from ovarian carcinoma 272 following cocultivation of the tumor cells with autologous PBMC in the presence of the 4 mAb combination. (b) Decreased expression of Gata3, increased expression of Tbx2 and IFNγ genes, and increased ratio of Tbx21/Gata3 by lymphoid cells which had been cocultivated with autologous tumor cells in the presence of the 4 mAb combination. (c) Increased number of CD3$^+$, CD4$^+$, and CD8$^+$ cells as well as CD44$^+$CD62L$^-$ memory T cells and CD11C$^+$CD83$^+$ mature dendritic cells and decreased number of CD19$^+$ B cells and CD4$^+$Foxp3$^+$ Tregs after cocultivation of 272 ovarian carcinoma cells with autologous PBMC in the presence of the 4 mAb combination

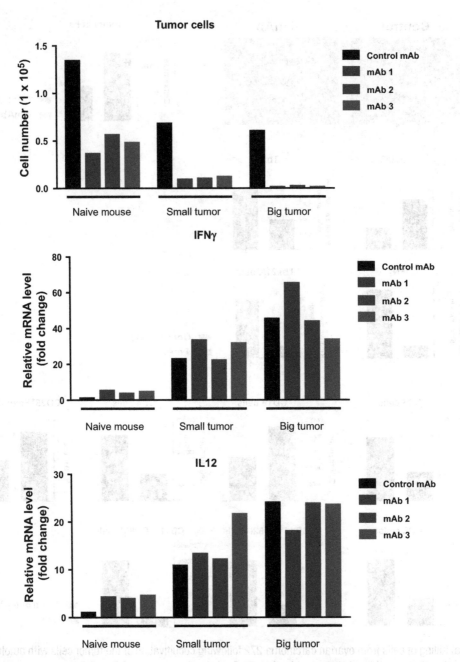

Fig. 7 (**a**) Shows the killing of mouse SW1 melanoma cells by in vitro sensitized spleen cells from a naïve mouse and from mice with a small or large tumor in the presence of each of three different immunostimulatory mAbs. Expression of the gene encoding IFNγ (**b**) or IL12 (**c**) correlates with the tumor cell killing data

therapy and the role of various lymphoid cell populations on the generation of a tumor-destructive Th1 response. Other efforts will be devoted to evaluating a variety of immunomodulatory mAbs, lymphokines and small molecules, alone and in combinations. Furthermore, autologous combinations of human PBMC and tumors will be cultured together with clinical grade immunomodulatory mAbs to investigate to what extent the in vitro sensitization model predicts patients' response to therapy. Other studies will investigate whether the anti-tumor response can be improved by combining immunomodulatory mAbs with immunization against epitopes that are overexpressed by the treated tumor.

Acknowledgments

We much appreciate technical contributions by Yuen Yee Yip, suggestions from Drs. Paul Abrams and Hans Olov Sjogren and clinical input from Drs. Elizabeth Swisher and John Liao, Department of Gynecology, University of Washington. The Hellstrom Foundation provided financial support.

References

1. Chen D, Mellman I (2013) Oncology meets immunology: the cancer immunity cycle. Immunity 39:1–10
2. Leach DR, Krummel MF, Allison JP (1996) Enhancement of antitumor immunity by CTLA-4 blockade. Science 271(5256):1734–1736
3. Melero I et al (1997) Monoclonal antibodies against the 4-1BB T-cell activation molecule eradicate established tumors. Nat Med 3(6):682–685
4. Pardoll D (2012) The blockade of immune checkpoints in cancer immunotherapy. Nat Rev Cancer 12:252–264
5. Dai M et al (2013) Long-lasting complete regression of established mouse tumors by counteracting Th2 inflammation. J Immunother 36:248–257
6. Dai M et al (2015) Curing mice with large tumors by locally delivering combinations of immunomodulatory antibodies. Clin Cancer Res 21(5):1127–1138
7. Hodi F et al (2010) Improved survival with ipilimumab in patients with metastatic melanoma. New Engl J Med 363:711–723
8. Ascierto P et al (2010) Clinical experiences with anti-CD137 and anti-PD1 therapeutic antibodies. Semin Oncol 37:508–516
9. Balachandran VP, Zeng S, Bamboat ZM, Ocuin LM, Obaid H, Sorenson EC, Popow R, Ariyan C, Rossi F, Besmer P, Guo T, Antonescu CR, Taguchi T, Yuan J, Wolchok JD, Allison JP, RP DM (2011) Imatinib potentiates antitumor T cell responses in gastrointestinal stromal tumor through the inhibition of Ido. Nat Med 17(9):1094–1100
10. Hamid O et al (2013) Safety and tumor responses with lambrolizumab (anti-PD-1) in melanoma. N Engl J Med 369(2):134–144
11. Wolchok JD et al (2013) Nivolumab plus ipilimumab in advanced melanoma. N Engl J Med 369(2):122–133
12. Galon J et al (2013) The continuum of cancer immunosurveillance: prognostic, predictive, and mechanistic signatures. Immunity 39:11–26
13. Couzin-Frankel J (2013) Cancer immunotherapy. Science 342:1432–1433
14. Forde PM et al (2018) Neoadjuvant PD-1 blockade in resectable lung cancer. N Engl J Med 378(21):1976–1986
15. Migden MR et al (2018) PD-1 blockade with cemiplimab in advanced cutaneous squamous-cell carcinoma. N Engl J Med 379(4):341–351
16. Di Giacomo AM, Biagioli M, Maio M (2010) The emerging toxicity profiles of anti-CTLA-4 antibodies across clinical indications. Semin Oncol 37(5):499–507
17. Bertrand A et al (2015) Immune related adverse events associated with anti-CTLA-4 antibodies: systematic review and meta-analysis. BMC Med 13:211
18. Postow MA, Sidlow R, Hellmann MD (2018) Immune-related adverse events associated with immune checkpoint blockade. N Engl J Med 378(2):158–168
19. Dai M et al (2016) An in vitro model that predicts the therapeutic efficacy of

immunomodulatory antibodies. J Immunother 39(8):298–305

20. Dai M et al (2018) Tumor regression and cure depends on sustained th1 responses. J Immunother 41(8):369–378

21. Levine BL et al (1997) Effects of CD28 costimulation on long-term proliferation of CD4+ T cells in the absence of exogenous feeder cells. J Immunol 159(12):5921–5930

22. Garlie NK et al (1999) T cells coactivated with immobilized anti-CD3 and anti-CD28 as potential immunotherapy for cancer. J Immunother 22(4):336–345

23. Mullen AC et al (2001) Role of T-bet in commitment of TH1 cells before IL-12-dependent selection. Science 292(5523):1907–1910

24. Minter LM et al (2005) Inhibitors of gamma-secretase block in vivo and in vitro T helper type 1 polarization by preventing Notch upregulation of Tbx21. Nat Immunol 6(7):680–688

Chapter 2

High-Throughput Direct Cell Counting Method for Immuno-Oncology Functional Assays Using Image Cytometry

Leo Li-Ying Chan

Abstract

The last decade has seen significant progress in the field of immuno-oncology. The increase in new technologies has led to the development of novel immuno-oncology assays to investigate immune responses to cancer cells. Numerous cell-based assays have utilized image cytometry due to its abilities to perform high-throughput screening, observe and automatically analyze cells directly in standard microplates, and digitally store the acquired cell images. In this chapter, we describe the use of Celigo Image Cytometer for immuno-oncology functional assays such as complement-dependent cytotoxicity (CDC), antibody-dependent cell-mediated cytotoxicity (ADCC), T cell proliferation, and T cell migration.

Key words Antibody drug conjugate (ADC), Complement-dependent cytotoxicity (CDC), Antibody-dependent cell-mediated cytotoxicity (ADCC), Direct cell-mediated cytotoxicity, Phagocytosis, Efferocytosis, T cell activation, T cell proliferation, T cell migration

1 Introduction

Cancer immunotherapy was named the breakthrough of the year in 2013 by *Science* magazine. Since then, the field of immuno-oncology has grown tremendously, ranging from basic research to clinical trials and from antibody drug conjugates (ADCs) to chimeric antigen receptor (CAR) T cell therapy. Numerous in vitro assays have been developed to investigate different cancer immunotherapies [1]. Traditionally, these are flow cytometry, luminescence, or release assays.

Co-culture-based cytotoxicity assays of immune (effector) and cancer (target) cells such as antibody-dependent cell-mediated cytotoxicity (ADCC) have been typically measured by release assays using Chromium^{51} (^{51}Cr), Calcein AM, or lactate dehydrogenase (LDH). In general, release assays require labeling cancer cells with these molecules or substances prior to co-culturing with the immune cells [2–4]. As the effector cells kill the target cells, these labels are released into the supernatant and measured by a plate reader for the level of radioactivity, fluorescence, or cytosolic enzyme, which can indicate the amount of cytotoxicity. Unfortunately, release assay is an indirect method for determining target cell death, requiring large amounts of target cells (>100K/well), and potential handling of hazardous material (^{51}Cr or ^{101}In). Other assays testing the efficacy of ADCs or complement-dependent cytotoxicity (CDC) have utilized flow cytometry to directly measure target cell viability after treatment [5, 6]. Similarly, flow cytometry requires large amounts of target cells that must be trypsinized for adherent cells, thus disturbing the natural state of the cells.

Many in vitro immunotherapy assays developed in recent years have shifted toward image-based platforms such as Opera Phenix (PerkinElmer, Waltham, MA, USA), Cytation (Biotek, Winooski, VT, USA) [7], IncuCyte (Essen Bioscience, Ann Arbor, MI, USA) [8], and Celigo Image Cytometer. Specifically, Celigo Image Cytometer (Nexcelom Bioscience, Lawrence, MA, USA) has been used in various in vitro immunotherapy assays in previous publications [9–12]. The Celigo employs a transmission and epifluorescence optical setup for one bright-field (BF) and four fluorescent (FL) imaging channels (blue, green, red, and far red) to perform plate-based image cytometric analyses. Both BF and FL imaging channels utilize high-power light-emitting diodes (LEDs) for illumination and excitation. Each FL imaging channel has a specific filter set for each color: blue (EX 377/50 nm, EM 470/22 nm), green (EX 483/32 nm, EM 536/40 nm), red (EX 531/40 nm, EM 629/53 nm), and far red (EX 628/40 nm, EM 688/31 nm). The combined optics and digital imaging allow variable imaging resolutions from 1 to 8 μm^2/pixel. The proprietary optics setup can capture highly uniform images of each well on a standard microplate. In addition, the F-Theta lens and Galvanometric mirrors enable a rapid image capturing process, where 96 whole-well BF images can be acquired in less than 4 min. The motion stage moves to the center of each well and rapidly captures 16 images using the Galvanometric mirrors, which allows minimum movement of the plate holder stage, and reduces the time required to scan the entire plate and limits sample perturbation. Cells seeded in standard microplates (6, 12, 24, 48, 96, 384, and 1536 well) can be autofocused in two ways using the Celigo software. Image-Based Auto-Focus (IBAF) uses a software algorithm to step up- and downward in the Z-axis to determine the best focus for each well. Hardware-Based Auto-Focus (HBAF) measures the distance between two beams indicating the thickness of the bottom plastic film, and the measured distanced is applied to each well to focus. In this work, Greiner 96-well microplates were used due to their consistency and flatness across the well and plate.

The Celigo software consists of five major steps START, SCAN, ANALYZE, GATE, and RESULTS. The START section allows users to select the type of microplate, as well as entering general information about the experiment. The SCAN section allows users to setup imaging and scanning parameters, such as imaging channels, exposure time, focusing method, and areas to scan on the plate. The ANALYZE section allows users to setup image analysis parameters to count cells of interest in the plate. The GATE section allows users to generate histograms and scatterplots of data from the captured images and perform a flow-like gating process. Finally, the RESULTS section allows the users to view counts, export data to EXCEL and FCS Express (De Novo Software, Glendale, CA, USA), and generate whole-well stitched images.

Image cytometric analysis methods can help eliminate issues faced by traditional in vitro functional immunotherapy assays. For example, directly counting cells in standard microplates requires much fewer cells (<10K/well) and does not require trypsinization for adherent cells. Furthermore, visually verifying effector and target cells in microplates can minimize uncertainties of immune response activities, as well as observation of changes in cell morphology. In addition, image cytometry can perform time-course monitoring of target and effector cells to further characterize immune response activities. Many in vitro immunotherapy assays have been migrating to imaging systems, and numerous approaches are being developed to streamline the entire research and development process for immuno-oncology. In this chapter, we will introduce various image-based immunotherapy functional assays utilizing the Celigo Image Cytometer.

2 2D Cell-Based Assays

Most in vitro immunotherapy assays utilize a co-culture model of immune (effector) and cancer (target) cells [2, 3, 13]. In general, the assays are performed with a monolayer of target cells seeded in a standard multi-well microplate with the addition of effector cells, antibodies, or small molecules. The purpose is to measure the live cell count or viability of the target cells to determine the level of cytotoxicity induced by the effector cells/antibody/complement. In this section, we will review 2D cell-based assays performed on the Celigo Image Cytometer for antibody drug conjugate, complement-dependent cytotoxicity, direct cell-mediated cytotoxicity, antibody-dependent cell-mediated cytotoxicity, bispecific antibody-based cytotoxicity, and phagocytosis and efferocytosis assays.

2.1 Antibody Drug Conjugate (ADC)

ADCs have been utilized to efficiently deliver cytotoxic agents directly to cancer cells through highly specific and targeted monoclonal antibodies [14]. To determine ADC efficacy, researchers can use image cytometry to measure cell count/confluence and viability of target cancer cells after treatment. For this assay, we demonstrate two methods to measure ADC potencies. In general, target cancer cells were first seeded in microplate wells with dose-dependent ADC treatments. Next, the Celigo was used to capture BF and/or FL images during treatment. Finally, the BF and FL images were analyzed to measure cell count/confluence and/or drug-treated target cell viability.

The first detection method was measuring changes in cell count or confluence over time in comparison to a negative control using BF images. In this experiment, the lung cancer cell line NCI-H1975 was seeded at 5,000 cells/well in a 96-well plate and treated with four concentrations of the tested ADCs. The Celigo Image Cytometer was used to acquire whole-well BF images and measure confluence percentages daily for 4 days.

The captured time-course BF images are shown in Fig. 1a. It is clear that the negative control became fully confluent while the positive treatment showed decreased confluence due to cell death. Well-by-well growth curves (Fig. 1b) directly plotted in the Celigo

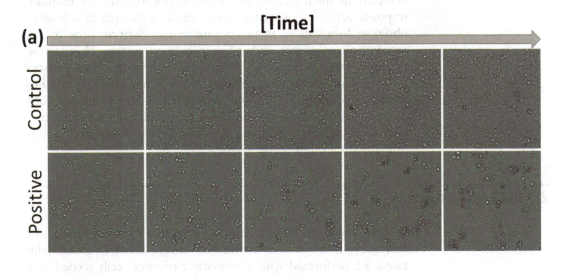

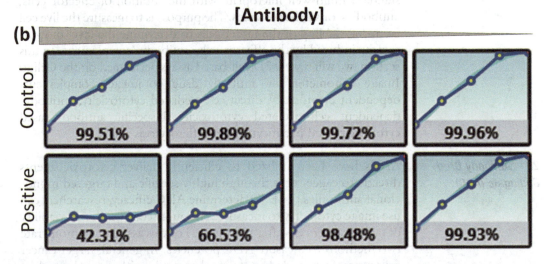

Fig. 1 BF-based imaging and analysis of ADC dose response. (**a**) Time-dependent BF images showing morphological changes to cells treated with positive ADC. (**b**) Celigo generated growth curves for different antibody concentrations

software showed changes in confluence between negative and positive controls at different antibody concentrations. The plate view in Fig. 2a shows the antibody dose-dependent coverage of target cells (pseudo-color green) on day 4. The negative control had fully confluent wells for each antibody concentration. In contrast, the positive control showed that high antibody concentration reduced cell coverage. The confluence percentage results are also plotted in Fig. 2b, showing corresponding dose-response data.

The second detection method was measuring treated target cell viability at a specific time point. In this experiment, 500 target cells were seeded per well in a 384-well plate and treated with nine concentrations of the tested ADCs. After 5 days, a combination

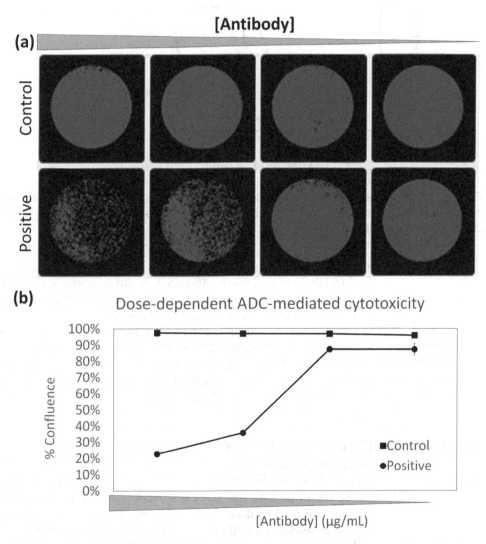

Fig. 2 ADC dose-dependent analysis. (**a**) Whole well view showing surface cell coverage (pseudo-color green) at different ADC concentrations. (**b**) ADC dose-dependent results showing high ADC effects with positive ADC

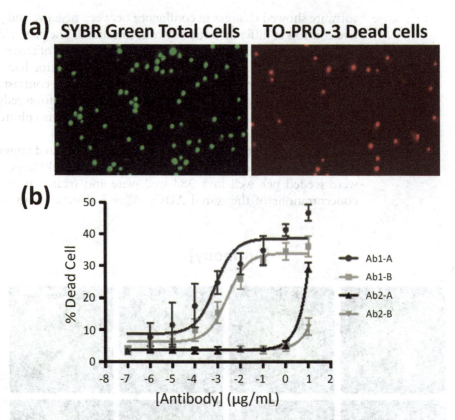

Fig. 3 FL-based imaging and analysis of the ADC dose response. (**a**) FL images of SYBR Green-stained total cells and TO-PRO-3-stained dead cells. (**b**) Measured dead cell percentages of four different ADCs showing clear dose responses for Ab1-A and Ab1-B

of SYBR Green (Total-Green) and TO-PRO-3 (Dead-Red) was used to label target cells for 45 min at room temperature. Finally, the green and red FL images were captured and analyzed to determine the dose-dependent percent dead results.

The SYBR Green and TO-PRO-3 FL images are shown in Fig. 3a. The green and red FL-positive cells were directly counted in the Celigo software to generate percentages of dead cells for each antibody concentration. The results showed obvious dose-response curves for two positive antibodies and low responses for the two negative antibodies (Fig. 3b).

2.2 Complement-Dependent Cytotoxicity (CDC)

The CDC assay is a method that allows researchers to measure direct complement killing of cancer cells using membrane attack complex [15]. As described above, two image cytometry methods can be performed. In the first experiment, target Daudi cells were stained with 5 µM calcein AM for 30 min and washed. Next, the cells were seeded 10,000 cells/well in a 96-well plate and treated with titrations of target antibodies and 2% human sera (complement). Finally, plates were scanned at 0, 30, 60, 90, and 120 min to count the number of live target cells (calcein⁺) and calculate cytotoxicity percentages over time. An equation (Eq. 1) was used to calculate cytotoxicity percentages using the number of calcein⁺ cells at each time point, where it represented the number of live target cells remaining in the well, thus referencing or normalizing on a per well basis. Furthermore, the negative control cytotoxicity percentage can be subtracted from experimental sample to normalize against baseline cell death to improve accuracy of the results.

$$\text{Cytotoxicity}\% = \frac{(\text{Calcein}_{t=0} - \text{Calcein}_{t=x})}{\text{Calcein}_{t=0}} \times 100 \qquad (1)$$

BF and calcein FL images for high and low antibody concentrations are shown in Fig. 4a. The low antibody concentration showed more calcein⁺ cells than the high concentration, where all of the cells have died and released the calcein. The cytotoxicity results are presented as time-dependent and dose-dependent plots (Fig. 4b, c). Over time, target cell cytotoxicity increased for the

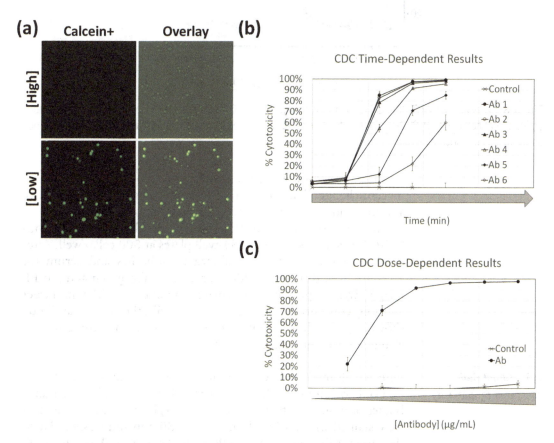

Fig. 4 Calcein-based imaging and analysis of CDC treatment. (**a**) FL and BF/FL overlay images of calcein-stained target cells treated with different concentrations of antibodies and complements. (**b**) CDC time-dependent results showing cytotoxic effects at different antibody concentrations. (**c**) CDC dose-dependent results showing an obvious dose response for the positive antibody

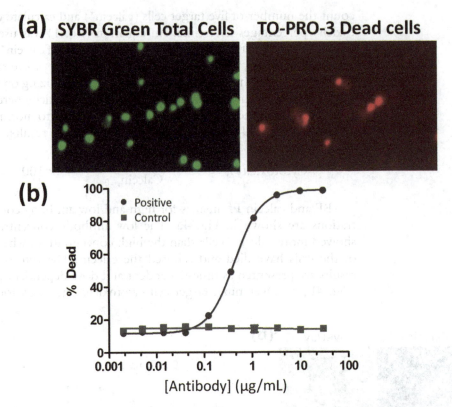

Fig. 5 SYBR Green and TO-PRO-3 imaging and analysis of CDC treatment. (**a**) FL images of SYBR Green-stained total cells and TO-PRO-3-stained dead cells. (**b**) Measured dead cell percentages of positive antibody showing a clear CDC dose response

positive antibodies, while the control condition did not show significant changes. The endpoint dose response showed that positive antibody induced a high CDC effect in target cells.

In the second experiment, target Daudi cells were stained with SYBR Green and seeded in 384-well plates at 500 cells/well. After incubation with a titration of target antibodies and serum for 15 min, TO-PRO-3 was added for 1 h. Finally, green and red FL images were captured to calculate viabilities at different target antibody concentrations (Fig. 5a). In Fig. 5b, the positive antibody showed a clear dose response for CDC effect, while there was no change in percent dead cells for the negative control.

2.3 Direct Cell-Mediated Cytotoxicity

Direct cell-mediated cytotoxicity assay measures cytotoxicity induced by immune cells on cancer cells without additional antibodies or other agents. In this experiment, K562 target cancer cells were stained with 5 μM calcein AM for 30 min and washed. Next, the target cells were seeded in 96-well plates at 10,000 cells/well. Subsequently, natural killer (NK) effector cells were added to the wells at effector-to-target (E:T) ratios of 10:1, 5:1, 2.5:1, 1.3:1, 0.6:1, and 0.3:1. Green FL images were captured, and live target cells (calcein[+]) were counted over time from 1 to 4 h. This allowed cytotoxicity percentages to be calculated by dividing cell counts for each sample with the negative control wells at different time points.

The time-dependent FL images in Fig. 6a showed fewer calcein⁺ cells over 4 h. In Fig. 6b, E:T ratio-dependent cytotoxicity percentages are plotted in respect to time, which increased with greater E:T ratios. The FL images and results are also presented as an endpoint E:T ratio-dependent plot shown in Fig. 7a, b.

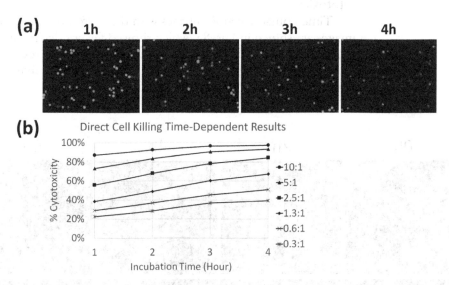

Fig. 6 Calcein-based imaging and analysis of direct cell-mediated cytotoxicity assay over time. (**a**) Time-dependent FL images showing a decrease in calcein⁺ target cells over time. (**b**) Time-dependent cytotoxicity % results showing different levels of cell killing with respect to E:T ratios

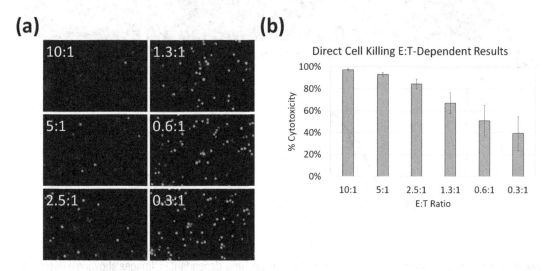

Fig. 7 Calcein-based imaging and analysis of direct cell-mediated cytotoxicity assay with respect to E:T ratios. (**a**) E:T ratio-dependent FL images showing a decrease in calcein⁺ target cells at 4 h. (**b**) E:T ratio-dependent cytotoxicity % results showing different levels of cell killing at 4 h

2.4 Antibody-Dependent Cell-Mediated Cytotoxicity (ADCC)

The ADCC assay quantifies cytotoxicity induced by immune cells on cancer cells with the addition of targeted antibody. In this experiment, MDA-MB-231 target cells were stained with 5 µM calcein AM for 30 min and seeded in the 96-well plate at 10,000 cells/well. Next, titrations of target antibodies and NK92 effector cells were added to the wells at a 10:1 E:T ratio. Finally, BF and green FL images were captured at time 0, 2, 4, and 6 h, where the number of calcein$^+$ target cells were counted to calculate cytotoxicity.

Time-course green FL images with the corresponding counted images are shown in Fig. 8a, which showed fewer calcein$^+$ cells due to ADCC. The calculated dose-dependent cytotoxicity results were plotted for each time point and revealed comparable killing between 4 and 6 h (Fig. 8b).

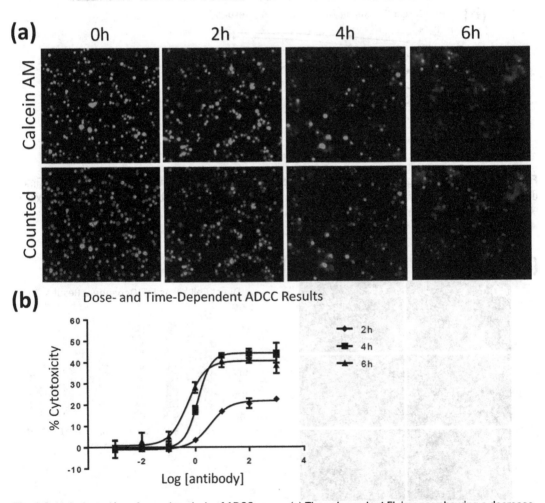

Fig. 8 Calcein-based imaging and analysis of ADCC assay. (**a**) Time-dependent FL images showing a decrease in calcein$^+$ target cells over 6 h. (**b**) Time-dependent cytotoxicity % results showing different levels of ADCC killing over time

2.5 Bispecific Antibody-Based Cytotoxicity

Bispecific antibody-based cytotoxicity assay requires co-labeling of two cell populations to measure the reduction in numbers of two target cell types. In this experiment, two target cell types (Parental NCI-H358 and NCI-H358 HER2 KO) were stained CellTracker™ Green CMFDA and CellTracker™ Violet BMQC, respectively [11]. They were seeded at 1:1 ratio with different bispecific antibodies and allowed to incubate for 3 days. On day 3, the cells were trypsinized, and propidium iodide was added to the wells to quantify the dead cells. The Celigo Image Cytometer was used to acquire BF and green, blue, and red FL images on day 3. The numbers of green, blue, and red fluorescent positive cells were counted to determine the cytotoxic effects of the bispecific antibodies.

Bispecific antibody-based cytotoxicity was determined by measuring cell viability on day 3. Different bispecific antibodies exerted various cytotoxic effects on parental and HER2 KO NCI-H358 cells. VkS93A + VHP97A/HER2 and VkF94A + VHP97A/HER2 DuetMab induced the highest differences in cell viability between the two cell types (Fig. 9).

2.6 Phagocytosis and Efferocytosis

Phagocytosis and efferocytosis assays are typically measured by flow cytometry [16], where macrophage and apoptotic/cancer cells are stained with Carboxyfluorescein succinimidyl ester (CFSE) and CellTrace Violet (CTV) dyes. The cells are analyzed to identify cell populations that are CFSE and CTV double positive to assess co-localization of the two cell types. Using image cytometry, cells can be left undisturbed in the wells to facilitate time-course visual monitoring of phagocytosis. In this experiment, macrophages were seeded into 96-well plates at 10,000 cells/well. Next, pHrodo-Red-labeled apoptotic cancer cells were added at a 10:1 E:T ratio. Fluorescent pHrodo label only fluoresces after phagocytosis due to lowering of pH, which can be quantified to measure efferocytosis. Two drugs were added to the wells to determine their phagocytosis inhibitory effects. After 48 h of incubation, the Celigo Image Cytometer was used to acquire BF and red FL images to count total pHrodo-Red positive cells.

The BF, red FL, and counted images are shown in Fig. 10a and showed reductions in pHrodo-Red positive cells due to the drug treatments. To compare the efferocytosis results using pHrodo, the integrated FL intensities were measured for each sample and compared to the negative control. The results showed reduced pHrodo fluorescence intensity, indicating inhibition of efferocytosis.

3 3D Tumor Spheroid-Based Assays

In recent years, tumor microenvironment has become a significant area of interest for immuno-oncology researchers seeking to

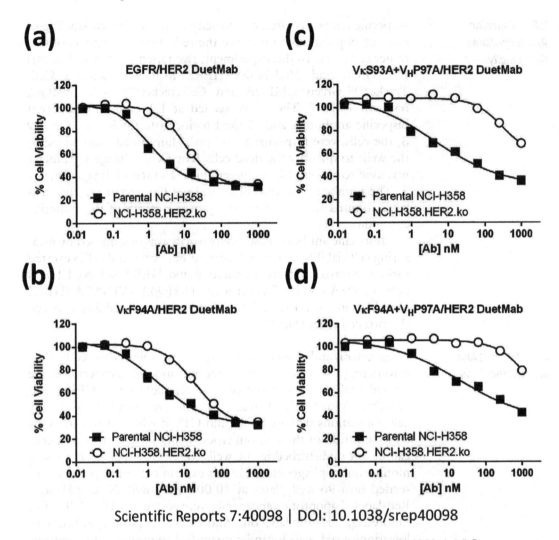

Fig. 9 Bispecific antibody killing assay results presented in Scientific Reports 7:40098. (**a–d**) Dose-response curves generated by measuring cell viabilities following different treatments with bispecific antibodies of both cell types

identify factors affecting immune cell response near tumor sites. Recent publications have also indicated that more researchers are constructing in vitro tumor microenvironment models such as co-cultures of 3D tumor spheroids, immune cells, and normal fibroblasts [17–19]. Using image cytometry, these co-cultures can be closely monitored and analyzed with BF and FL imaging.

3.1 Antibody-Dependent Cell-Mediated Cytotoxicity (ADCC)

ADCC assays can be performed using 3D tumor spheroid models by introducing immune effector cells to an in vitro system. We demonstrated the use of ImmTAC molecules (Immunocore, Milton Park, Abingdon, Oxfordshire, UK) that can recognize and bind to cancer cells expressing peptide-human leukocyte antigen (HLA)

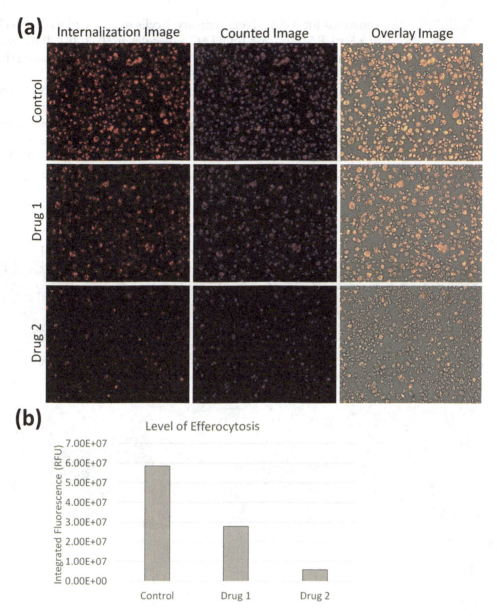

Fig. 10 Efferocytosis imaging and analysis using pHrodo staining. (**a**) FL and BF/FL overlay images showing reduction in efferocytosis with drug treatment. (**b**) Integrated FL intensity measurement showing the level of efferocytosis

targets. The circulating T cells are recruited to the tumor site by interacting with the anti-CD3 fragment-free end of the ImmTAC molecule. Then, T cells are activated to release cancer cell lytic granules to induce cytotoxicity.

In this experiment, GFP-expressing MDA-MB-453 target cells were first seeded in an ultra-low attachment U-bottom plate at 1,000 cells/well. The plate was centrifuged and allowed to form spheroids for 3 days. Next, primary T cells were added to each well at a 50:1 E:T ratio and ImmTAC molecules at 10, 1, 0.1, 0.01, and 0.001 nM. Finally, BF and green FL images were acquired using the Celigo Image Cytometer on days 1 and 5.

The BF and green FL overlay images of the spheroids are presented in Fig. 11a, which showed decreased GFP area with respect to increased ImmTAC molecule concentration. The spheroid GFP area results are shown in Fig. 11b, where a clear dose response is observed. The positive control with 20% DMSO showed a significant reduction in GFP area similar to the highest dosage of ImmTAC.

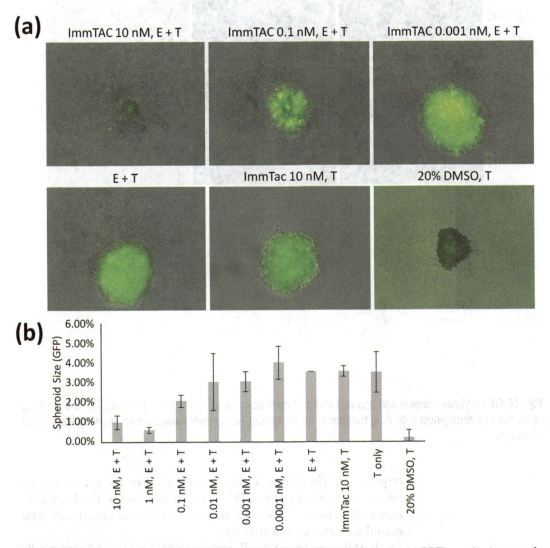

Fig. 11 ADCC assay analysis of 3D tumor spheroid model co-cultured with T cells. (**a**) BF/FL overlay images of GFP[+] tumor spheroid with decreasing size due to T cell/ImmTAC killing. (**b**) Spheroid size measurement by GFP confluence % showing the dose response with respect to ImmTAC concentration

4 T Cell and CAR T Cell-Based Assays

In the recent years, CAR T cell-based therapy has become the focus of attention for the next breakthrough in cancer immunotherapy. This is largely due to the approval of two CAR T cell therapies: tisagenlecleucel (Kymriah) from Novartis and Axicabtagene ciloleucel (Yescarta) from Kite Pharma/Gilead [20]. Bringing a CAR T cell treatment to the market required a tremendous amount of time and effort from research and development to manufacturing, not to mention clinical trials. In this section, we introduce cell-based assays that investigate the cytotoxicity capabilities of CAR T cells.

4.1 T Cell Activation and Proliferation Measurement

Two important factors to consider when measuring T cell or CAR T cell killing capability are activation and proliferation. Traditionally, these can be measured in T cells by flow cytometry using CFSE staining or enzyme-linked immunosorbent assays (ELISAs) for cytokine release. However, CFSE staining can interfere with T cell activities, and ELISAs require manipulation of cells that may affect their natural state. Image cytometry methods enable direct counting of proliferating T cells in the well, thus providing time-course monitoring of T cell activities. In this experiment, T cells were seeded into 96-well plates at 5,000 cells/well in the presence or absence of CD3 and CD28 for T cell activation. Next, they were allowed to incubate for 48 h, and BF images were captured at 0, 24, and 48 h. The total activated T cells were directly counted from BF images to demonstrate time-course monitoring of T cell activation and proliferation.

Using the gating function of Celigo software, the number of larger activated T cells were directly counted from day 0 to 3 (Fig. 12a). Notably, unstimulated T cells did not increase in size visually or on the size histogram. The T cell counting results in Fig. 12b showed an increase in proliferation of activated T cells from 5,000 to approximately 150,000 cells after 48 h of stimulation.

4.2 T Cell Migration

Another important factor in measuring T cell or CAR T killing capability is investigating the ability of stimulated T cells to migrate to tumor sites and kill the target cells. In this experiment, T cells were seeded in the Transwell inserts on top of 96-well plates. A mixture of chemokines (0–100 nM) and 8 μM Hoechst staining solution were introduced to the bottom wells to induce T cell migration through the Transwell insert pores. Since T cells are suspension cells, once they migrate through the pores, they fall to the bottom where they are stained with Hoechst. After allowing 4 h for migration, BF and blue FL images were acquired, and total migrated cells were directly counted for each well.

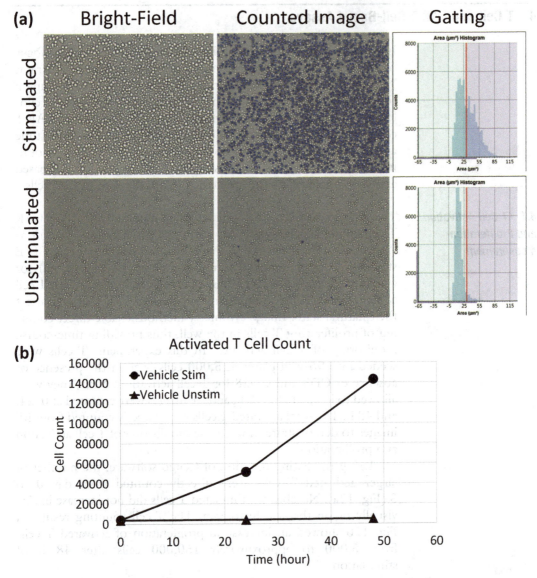

Fig. 12 T cell activation and proliferation analysis using BF imaging. (**a**) BF images showing direct cell counting of activated T cells using the size gating function of Celigo software. (**b**) Time-dependent T cell counting results showing an increase in activated T cells over 48 h

An example of the whole-well and zoomed-in BF and Hoechst FL images are shown in Fig. 13a. The whole plate view counts are provided in Fig. 13b, which provided the number of migrated T cells with respect to chemokine concentrations; the mean results are plotted in Fig. 13c. The BF images can also be counted directly (Fig. 13d). Interestingly, only the middle concentration (1 nM) achieved the highest T cell migration, which suggested that low concentrations were insufficient to induce migration, while high concentrations potentially exhausted the cells and inhibited migration capability.

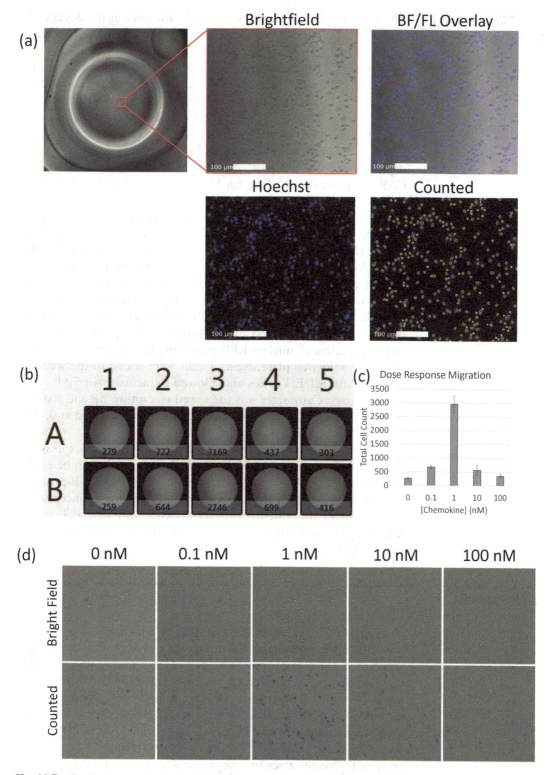

Fig. 13 T cell migration analysis using BF and FL imaging. (**a**) BF and FL images showing direct cell counting of T cells in the bottom of a Transwell. (**b**) Whole plate view showing direct cell counting results. (**c**) Cell counts are plotted with maximum migration at 1 nM. (**d**) Example of direct cell counting using BF images

4.3 T Cell and CAR T Cell Killing Assay

The CAR T cell killing assay is essentially the same as the direct cell-mediated cytotoxicity assay. In this experiment, calcein-stained DAOY target cells were seeded in a 96-well plate at 10,000 cells/well. Next, CAR T cells with HER2 or CD19 CAR were added to each well at 40:1 E:T ratio. Finally, the co-cultures were allowed to incubate for 4 h before image cytometry analysis using Celigo.

The time-dependent BF and green FL images are shown in Fig. 14a, which showed decrease in number of calcein$^+$ DAOY cells. Cytotoxicity percentages were calculated using Eq. 1, and the E:T-dependent results are plotted in Fig. 14b, which showed higher CAR T cell-mediated killing for HER2 CAR T compared to CD19 CAR T.

T cell killing typically requires more than 24 h to measure cytotoxic effects, which requires FL tracking labels that persist longer than the calcein AM stain. The most effective method to detect live target cells over 24 h is the use of nuclear fluorescent protein-expressing cells. As the target cells die, fluorescent proteins can leak out and significantly reduce the fluorescence intensity of the cells. Therefore, live target cells with bright nuclear fluorescence can be easily counted and tracked over a long co-culture time. In this experiment, nuclear RFP-expressing target cancer cells were seeded in a 96-well plate. Next, T cells were added to the wells at 1:1, 2:1, and 5:1 E:T ratios and allowed to incubate for 50 h. The Celigo Image Cytometer was then used to capture BF and red FL images, where live RFP-positive target cells were counted at 0, 22, 26.5, 46, and 50.5 h.

Figure 15a shows an example of the RFP nuclear-expressing target cells from time 0 to 72 h, in which clear nuclei can be easily counted from the FL images. The average number of RFP-positive target cells are plotted with respect to time showing an increase in target cells in the absence of T cells, while different E:T ratios showed growth inhibition (1:1 E:T ratio) and cytotoxicity (2:1 and 5:1 E:T ratios).

5 Discussion

The purpose of in vitro immunotherapy functional assays is to investigate the potencies of immune cells and antibodies on targeted cancers using 2D or 3D models. Traditional assays involving flow cytometry, luminescence, and release assays are commonly used; however, they require significant numbers of cancer and immune cells (>100K and >1 million cells, respectively), which is highly demanding for researchers using precious primary samples. Secondly, these assays often indirectly measure only the supernatant; not directly assessing target cells can lead to uncertainty. Utilizing ^{51}Cr can also pose hazardous and financial concerns. Finally, researchers are required to trypsinize cells off microplates

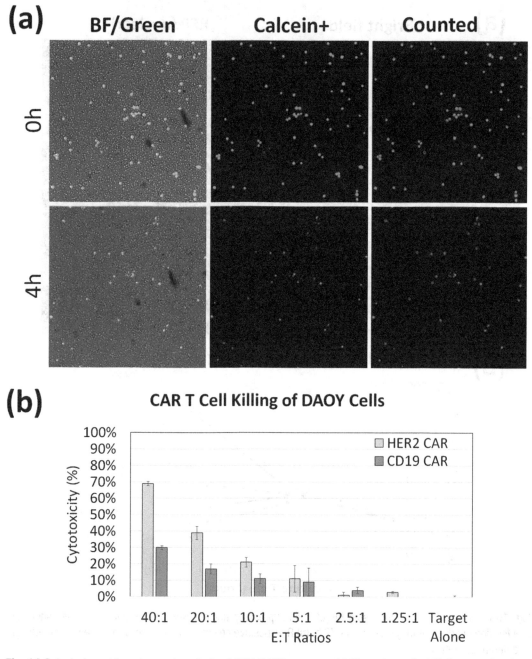

Fig. 14 Calcein-based imaging and analysis of CAR T killing assay. (**a**) Time-dependent FL images showing decreased numbers of calcein⁺ target cells over 4 h. (**b**) Comparison between HER2 and CD19 CAR T showing higher killing capacity for HER2 CAR T on DAOY target cells

to perform flow cytometry assays, which disturbs the natural state of the target cells.

To overcome these issues, many researchers have migrated toward image-based platforms for in vitro immunotherapy functional assays [9, 10, 21, 22]. Employing image cytometers such

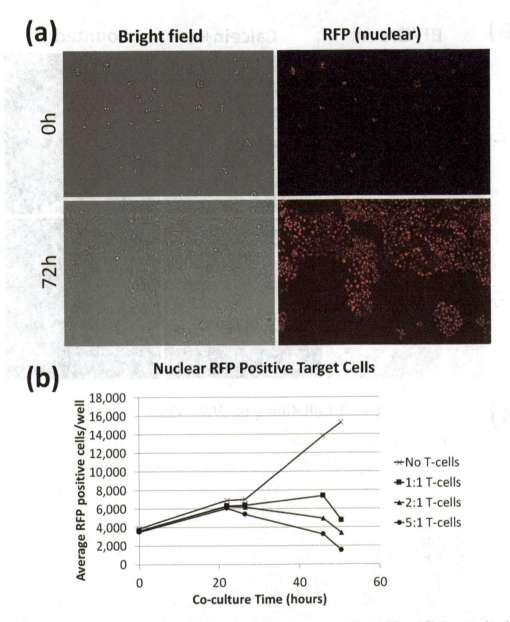

Fig. 15 Long-term T cell killing using target cells expressing nuclear RFP. (**a**) BF and FL images showing proliferation of adherent target cells over 72 h. (**b**) RFP-positive cell count results showing T cell killing effects at different E:T ratios

Celigo can eliminate the challenges posed by the traditional methods. Directly imaging and analyzing cells in standard microplates eliminates the need to trypsinize cells out of the plate. In addition, the number of cells required for image cytometry are 10–20× less than flow cytometer and release assays, which is advantageous for researchers working with primary samples. Furthermore, the ability to visualize and analyze cells in the wells enables the collection of more data related to BF and FL imaging, cell counting, and cell morphology, which can better characterize the tested cell samples.

The three main advantages of employing Celigo Image Cytometer are the ability to rapidly acquire and analyze cells in plates, generate uniform BF and 4-color FL images, and perform 2D and 3D high-throughput screening. Due to the speed, flexibility, and ease of use of the Celigo, the in vitro immunotherapy assays described in the previous sections have been developed and are currently performed by many immuno-oncology researchers.

Short- and long-term monoculture-based cytotoxicity assays can be easily performed on the Celigo Image Cytometer. Similarly, short-term co-culture cytotoxicity assays utilizing calcein AM to monitor and count live target cells over time has been rigorously developed. Long term co-culture cytotoxicity assays require tracking dyes that must remain brightly fluorescent for at least 48 h. Unfortunately, not many tracking dyes have that capability, thus the most optimized method would be to utilize nuclear or cytoplasmic fluorescent protein-expressing target cells to track the number of live cells over time. Future research challenges are identifying a novel and robust image cytometry method to monitor and analyze co-culture assays if primary target and immune cells are used. Similarly, further researches are required to develop and investigate assays related to tumor microenvironment, such as co-culturing immune and healthy cells, tumor spheroids, as well as the addition of other antibodies, small molecules, and check-point inhibitors.

The field of immuno-oncology has grown rapidly with novel cancer treatments discovered each day. Using newly available technologies such as image cytometers can improve efficiencies for performing current immunotherapy assays and lead to the development of new assays and methods to answer more complex questions.

Acknowledgments

The author would like to thank the external scientists of Nexcelom Bioscience LLC. for providing immuno-oncology experimental results from the Celigo Image Cytometer.

Conflicts of Interest: The author, L.L.C., declares competing financial interests. The methods and experiments presented in this book chapter were developed and performed using the Celigo Image Cytometer, which is a product of Nexcelom Bioscience, L.L.C. The experiments were performed to demonstrate novel in vitro immunotherapy functional assays that significantly improve upon assays currently performed in the field of immuno-oncology.

References

1. Couzin-Frankel J (2013) Cancer immunotherapy. Science 342:1432–1433
2. Li M, Zheng H, Duan Z, Liu H, Hu D, Bode A, Dong Z, Cao Y (2012) Promotion of cell proliferation and inhibition of ADCC by cancerous immunoglobulin expressed in cancer cell lines. Cell Mol Immunol 9:54–61
3. Somanchi SS, McCulley KJ, Somanchi A, Chan LL, Lee DA (2015) A novel method for assessment of natural killer cell cytotoxicity using image cytometry. PLoS One 10:e0141074
4. Zaritskaya L, Shurin MR, Sayers TJ, Malyguine AM (2010) New flow cytometric assays for monitoring cell-mediated cytotoxicity. Expert Rev Vaccines 9:601–616
5. Lindorfer MA, Cook EM, Tupitza JC, Zent CS, Burack R, Jong RNd, Beurskens FJ, Schuurman J, Parren PWHI, Taylor RP (2016) Real-time analysis of the detailed sequence of cellular events in mAb-mediated complement-dependent cytotoxicity of B-cell lines and of chronic lymphocytic leukemia B-cells. Mol Immunol 70:13–23
6. Riedl T, Boxtel Ev, Bosch M, Parren PWHI, Gerritsen AF (2016) High-throughput screening for internalizing antibodies by homogeneous fluorescence imaging of a pH-activated probe. J Biomol Screen 21:12–23
7. Fassy J, Tsalkitzi K, Goncalves-Maia M, Braud VM (2017) A real-time cytotoxicity assay as an alternative to the standard chromium-51 release assay for measurement of human NK and T cell cytotoxic activity. Curr Protoc Immunol 118:7.42.1–7.42.12
8. Cerignoli F, Abassi YA, Lamarche BJ, Guenther G, Ana DS, Guimet D, Zhang W, Zhang J, Xi B (2018) In vitro immunotherapy potency assays using real-time cell analysis. PLoS One 13:e0193498
9. David JM, Dominguez C, McCampbell KK, Gulley JL, Schlom J, Palena C (2017) A novel bifunctional anti-PD-L1/TGF-β Trap fusion protein (M7824) efficiently reverts mesenchymalization of human lung cancer cells. OncoImmunology 6:e1349589
10. Fantini M, David JM, Saric O, Dubeykovskiy A, Cui Y, Mavroukakis SA, Bristol A, Annunziata CM, Tsang KY, Arlen PM (2018) Preclinical characterization of a novel monoclonal antibody NEO-201 for the treatment of human carcinomas. Front Immunol 8:1899
11. Mazor Y, Sachsenmeier KF, Yang C, Hansen A, Filderman J, Mulgrew K, Wu H, Dall'Acqua WF (2017) Enhanced tumor-targeting selectivity by modulating bispecific antibody binding affinity and format valence. Sci Rep 7:Article ID 40098
12. Mazor Y, Yang C, Borrok MJ, Ayriss J, Aherne K, Wu H, Dall'Acqua WF (2016) Enhancement of immune effector functions by modulating IgG's intrinsic affinity for target antigen. PLoS One 11:e0157788
13. Clay TM, Hobeika AC, Mosca PJ, Lyerly HK, Morse MA (2001) Assays for monitoring cellular immune responses to active immunotherapy of cancer. Clin Cancer Res 7:1127–1135
14. Sievers EL, Senter PD (2013) Antibody-drug conjugates in cancer therapy. Annu Rev Med 64:15–29
15. Duensing TD, Watson SR (2018) Complement-dependent cytotoxicity assay. Cold Spring Harb Protoc 2018(2). https://doi.org/10.1101/pdb.prot093799
16. Tseng D, Volkmer J-P, Willingham SB, Contreras-Trujillo H, Fathman JW, Fernhoff NB, Seita J, Inlay MA, Weiskopf K, Miyanishi M, Weissman IL (2013) Anti-CD47 antibody-mediated phagocytosis of cancer by macrophages primes an effective antitumor T-cell response. Proc Natl Acad Sci U S A 110:11103–11108. https://doi.org/10.1073/pnas.1305569110
17. Hirschhaeuser F, Menne H, Dittfeld C, West J, Mueller-Klieser W, Kunz-Schughart A (2010) Multicellular tumor spheroids: an underestimated tool is catching up again. J Biotechnol 148:3–15
18. Kunz-Schughart LA, Freyer JP, Hofstaedter F, Ebner R (2004) The use of 3-D cultures for high-throughput screening: the multicellular spheroid model. SLAS Discov 9:273–285
19. Thoma CR, Zimmermann M, Agarkova I, Kelm J, Krek W (2014) 3D cell culture systems modeling tumor growth determinants in cancer target discovery. Adv Drug Deliv Rev 69-70:29–41
20. Yip A, Webster RM (2018) The market for chimeric antigen receptor T cell therapies. Nat Rev Drug Discov 17:161–162
21. Chan LL-Y, Smith T, Kumph KA, Kuksin D, Kessel S, Déry O, Cribbes S, Lai N, Qiu J (2016) A high-throughput AO/PI-based cell concentration and viability detection method using the Celigo image cytometry. Cytotechnology 68:2015–2025
22. Zhang H, Chan LL-Y, Rice W, Kassam N, Longhi MS, Zhao H, Robson SC, Gao W, Wu Y (2017) Novel high-throughput cell-based hybridoma screening methodology using the Celigo Image Cytometer. J Immunol Methods 447:23–30

Chapter 3

In Vitro Functional Assay Using Real-Time Cell Analysis for Assessing Cancer Immunotherapeutic Agents

Biao Xi, Peifang Ye, Vita Golubovskaya, and Yama Abassi

Abstract

Immuno-oncology is undoubtedly one of the fastest developing fields in cancer therapy, most recently propelled by successful clinical applications of chimeric antigen receptor (CAR)-based T cell therapy and monoclonal antibodies targeting immune checkpoints, namely, CTLA4, PD-1, and PD-L1. However, while the advancement of the cancer immunotherapy has brought tremendous hope to some patients, there is a pressing need for more predictive in vitro and in vivo translational models to aid in the development of novel cancer immunotherapeutic agents. As a potential solution, the xCELLigence Real-Time Cellular Analyzer (RTCA) system (ACEA Biosciences) was developed as a high-throughput, in vitro platform to monitor and screen the ability of pharmacological agents to modulate the cytolytic activity of the immune cells (i.e., the effector cell) against tumor cells (i.e., the target cell) in a real-time manner. This specialized microtiter plate system uses integrated gold microelectrodes as a sensor to measure the electronic current flow that is impeded by the tumor cells attached on the sensor surface. This measured impedance signal is dictated by cell number, size, and surface attachment strength of the adherent cells. Thus, under specific assay conditions, the xCELLigence RTCA is able to distinguish between the adherent target cancer cells and nonadherent effector immune cells. The RTCA platform has also been successfully adapted to monitor suspension target cells originated from hematological cancers. In this chapter, we outline several workflows for using the xCELLigence RTCA to evaluate the potency of clinically relevant cancer immunotherapeutic agents in modulating immune cell-mediated killing of hematological and solid tumor cells.

Key words Bispecific antibody, Cancer immunotherapy, Cell-based assay, Chimeric antigen receptor (CAR), Immuno-oncology, NK92, Real-time cellular analysis

1 Introduction

Several in vitro assays have been the staple for measuring cell proliferation, testing cytotoxic effects of pharmacological compounds, and multiplexing as an internal control to determine viable cell number during other cell-based assays. These include annexin V and propidium iodide staining by flow cytometry [1–3], MTT and ATP assays [4–6], chromium release assay [7, 8], and clonogenic assays [9–11]. Phenotypic assays are also available for

Seng-Lai Tan (ed.), *Immuno-Oncology: Cellular and Translational Approaches*, Methods in Pharmacology and Toxicology, https://doi.org/10.1007/978-1-0716-0171-6_3, © Springer Science+Business Media, LLC, part of Springer Nature 2020

monitoring and quantifying cellular immune responses against tumor cells, such as peptide MHC tetramers and TCR regions usage [12–14], lymphocyte proliferation assay [15–17], ELISPOT and flow cytometry-based measurement of specific cytokine-secreting cell populations [18–22], and quantification of cytotoxic lymphocytes by LDA [23]. More recently, direct and real-time measurement of the cytotoxicity of immune cells upon coculture with target tumor cells has increasingly been incorporated in the cancer immunotherapy field [24–28].

The xCELLigence Real-Time Cellular Analyzer (RTCA) (ACEA Biosciences) is a real-time and label-free technology platform developed more than a decade ago and has been widely applied to various cell-based functional assays, including cell quality control and cell therapy applications [25, 29–34]. The RTCA system has been successfully used to assess the cytolytic potency of cellular and biologic therapies against cancer cells, including but not limited to NK cells, CAR-T cells, therapeutic antibodies, and oncolytic virus [35–41].

The core of the xCELLigence technology is based on microelectronic sensors embedded at the bottom of each well of a microtiter plate, or an E-plate, designed to measure the impedance of the microelectronic current circulating inside the well, when media and cells are present. The measured impedance value is converted into a parameter named "Cell Index," which is determined by the extent of the cells covering the sensors and impeding the current flow (Fig. 1). Thus, the xCELLigence RTCA platform is extremely sensitive and has been applied to measure subtle morphological

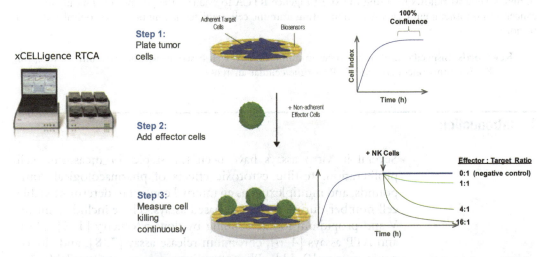

Fig. 1 Simple workflow and walk-away operation with the xCELLigence Real-Time Cell Analyzer (RTCA) System. In the RTCA assay, the target cells are seeded in the E-plate and allowed to attach and grow overnight. The effector cells (e.g. NK cells) are added into the same well. Cytotoxicity of the target cells are measured as progressive decrease in Cell Index as defined in the text

change of the cells in response to pathway activation [42–44], as well as contraction of cardiomyocytes [45, 46].

Cell Index values change in relation to cell number, size, and adherence strength and are measured at programmable intervals. The xCELLigence RTCA software displays Cell Index value over time or other related parameters like percentage of cytolysis of the target cells, inhibition concentration 50 (IC50), or the slope of the Cell Index curve.

Changes in impedance are reported as Cell Index (CI) and Normalized Cell Index (NCI), which have been described previously [26, 30, 33, 47, 48]. After normalizing the data to account for "target cell alone" and "effector cell alone" controls, parameters such as % Cytolysis and KT50 are determined using the xCELLigence RTCA software (ACEA Biosciences). The % Cytolysis plot utilizes a Target Alone Control and a Normalization Time to calculate the % of Cytolysis at every time point. For each well, the % Cytolysis utilizes the Normalized Sample Cell Index and the Normalized Average Target Alone Control according to the following equation:

$$\% \cdot Cytolysis = \left[1 - \frac{CI_{ti}/CI_{nml\_time}}{\left(CI_{Target\ Alone \cdot ti}/CI_{Target\ Alone \cdot nml\_time}\right)}\right] \times 100$$

where CI_{ti} is the average Cell Index between replicate wells at the time t_i, CI_{nml_time} is the average Cell Index of replicate wells at normalization time, $CI_{Target\ Alone\ ti}$ is the average Cell Index of replicate target alone control wells at the time t_i, and $CI_{Target\ Alone\ nml\_time}$ is the average Cell Index of replicate target alone control wells at normalization time.

Here, we outline a general workflow for applying the xCELLigence RTCA system to evaluate the killing potential of immunotherapeutic agents against solid tumor and liquid tumor cells. While this chapter mainly focuses on basic assay principle and protocol, the reader is referred to the xCELLigence manufacturer's instructions and customized software for more detailed protocols and analyses specific to each application.

2 Materials

1. NK92: Natural killer cell line derived from a non-Hodgkin's lymphoma patient, ATCC® CRL-2407™.

2. Raji: B lymphocytes cell line derived from a Burkitt's lymphoma patient, ATCC® CCL-86™.

3. BxPC3: Pancreas cancer cell lines derived from patient, adherent, ATCC® CRL-1687™.

4. Daudi: B lymphoblast cells from a Burkitt's lymphoma patient, ATCC® CCL-213™.
5. EGFR-CD28-CAR-T: provided by ProMab, PM-CAR-1021.
6. Mock-CAR-T: provided by ProMab, PM-CAR1000.
7. Non transduced T cells for CAR-T control: from ProMab, PM-CAR2003.
8. PBMC: iXCell Biotechnologies, 10HU-003.
9. FACS buffer: ProMab, 1× PBS (Ca/Mg, −/−) +0.5% BSA.
10. PBS (Ca/Mg, −/−): HyClone™ Phosphate Buffered Saline (PBS), SH30256.01.
11. FBS: HyClone™ Fetal Bovine Serum (US), Characterized, Fisher Scientific, SH300071.03.
12. RPMI 1640: HyClone™ RPMI 1640 Media, Fisher Scientific, SH3002701.
13. Ficoll-Paque Plus: GE Healthcare, 17-1440-02.
14. RBC Lysis Buffer (10×): Red Blood Cell Lysis buffer from Biolegend, 420301.
15. NK92 cell medium: Alpha Minimum Essential medium without ribonucleosides and deoxyribonucleosides but with 2 mM L-glutamine (Sigma G7513) and 1.5 g/L sodium bicarbonate (Sigma, M4526).

    To make the complete growth medium, the following components are added: 0.2 mM inositol (Sigma, I7508), 0.1 mM 2-mercaptoethanol (Sigma, M6250), 0.02 mM folic acid (Sigma, F8758), 12.5% horse serum (Fisher Scientific, 16050130), and 12.5% fetal bovine serum (Fisher Scientific, SH300071.03).
16. IL-2: PeproTech, 200-02. Prepare the stock at a high concentration and make aliquots. IL-2 is used at a concentration of 200 U/mL during cell culture but is not used in the functional assay.
17. 7-AAD: ThermoFisher, A1310.
18. Anti-CD19 hIgG1: G&P Bioscience, Cat. No. MAB0776.
19. CD3/CD19-BiTE: scFv(CD19xCD3e), G&P Bioscience, Cat. No. FCL1770.
20. RTCA: ACEA Biosciences, 00380601040.
21. Flow cytometry: ACEA Biosciences (PLEASE PROVIDE MODEL).
22. E Plate 96 View: ACEA Biosciences, 0647248001.
23. B cell killing kit (CD40): ACEA Biosciences, 8100004.
24. Human serum: GemCell US Origin Human Serum AB, 100-512.

25. T cell Enrichment kit, STEMCELL, EasySep™, Cat. No. 28741PIS.

3 Assessing NK92-Cell-Mediated Cytolytic Potency

NK92, a natural killer (NK) cell line derived from peripheral blood mononuclear cells of a 50-year-old Caucasian male with rapidly progressive non-Hodgkin's lymphoma, is a commonly used effector cell line for assessing NK cell-mediated cytotoxicity including antibody-dependent cellular cytotoxicity (ADCC). This cell line is maintained according to ATCC standard reagents and procedure. IL-2 is stored in 4 °C and should be freshly added into the medium with cells (not premixed with medium). In the example below, we utilize Raji as target cancer cells. As with all assays, the cell seeding density and effector to target cell ratio will need to determined and optimized empirically. Note that a single 96-well E-plate is generally adequate for more than dozens of treatment conditions.

1. Keep Raji cells and NK92 cells cultured for at least two passages if cells have been revived from storage in liquid nitrogen. NK92 cells could be used for assay within 3 months if the culture density is maintained less than 1,000,000 cells/mL.

2. Prepare the coating reagent from the B cell killing CD40 kit. Dilute Tethering Buffer, provided as 10 mL of 10× solution, by adding 90 mL of sterile tissue culture grade water to the bottle. Dilute tethering reagent (anti-CD40), 500 µg/mL, to the final concentration of 4 µg/mL to make the tethering reagent solution using diluted tethering buffer. Coat the E-plate View 96 by adding 50 µL of the 4 µg/mL tethering reagent solution to each well.

3. Incubate the E-plate at room temperature or 37 °C for 3 h. Then remove the entire coating buffer and wash the wells with diluted tethering buffer twice. Keep the E-plate inside the cell culture hood with its lid on to keep it from becoming completely dry.

   If the E-plate is not used immediately, keep coating buffer inside the plate and seal the plate inside 4 °C refrigerator for up to 2 weeks. Remove the coating buffer and wash the plate with tethering buffer before seeding cells.

4. Add 50 µL of RPMI 1640 medium to each well of the E-plate. Put the E-plate to the RTCA inside the 37 °C cell culture incubator for background reading, which is the impedance value of the medium only for each well. This takes about 1 min. Once finished, remove the E-plate from the RTCA and bring it to cell culture hood to add cells.

5. Transfer enough actively growing Raji cells from culture using a serological pipette to a 15 mL falcon tube. Add more medium to bring up the volume of 10–15 mL. Then centrifuge the cells at 300 × g for 5 min. Discard the supernatant and resuspend the cell pellet with 5 mL of fresh medium by serological pipetting gently. Count the cell number twice and use the average value as the cell density. Dilute the cells to the 600,000 cells/mL.

   Tip: To keep the consistency of the assay performance, avoid switching methods for counting cells between experiments. Make sure the cell suspension is homogeneous before sampling for cell counting.

6. Transfer diluted cells to a 10 mL reservoir and make sure cells are mixed well and then quickly transfer 100 µL of cell suspension to the E-plate with a multichannel pipette to seed 60,000 cells per well. Leave at least six wells for "no target cell" control wells.

7. Cover E-plate with lid and leave the E-plate inside the cell culture hood under room temperature for about 30 min. This is a very important step and should not be skipped.

8. Once the Raji cells are all settled onto bottom of the each well (verify under the microscope, if necessary), take the E-plate back to the RTCA.

9. Engage the plate in the RTCA to start monitoring the impedance by using the preset program to measure every 15 min or set time interval of your choice. Choose to display the real-time analysis of the Cell Index of Raji cells from all wells, which displays the cell index dynamics of the cell growth over time.

10. On the following day, prepare NK92 cells in fresh medium as described in **step 5** above using RMPI 1640 medium (without IL-2) to resuspend the NK92 cell pellet in about 5 mL. Make sure to prepare NK92 cells as a nice and even suspension by pipetting up and down gently with a serological pipette. Count the cells and prepare the cells at the density of 600,000 cells/mL with 37 °C warmed up RPMI 1640 full medium (without IL-2).

11. Prepare serial dilutions of the NK92 cell suspension in a multichannel reservoir with warmed-up RPMI 1640 full medium (without IL-2) at the densities of 600,000, 450,000, 300,000, 150,000, 60,000 cells/mL. This is calculated based on adding 100 µL of each diluted group of NK92 cells per well, which will result in NK92:Raji cell ratios or Effector–Target cell (E:T) ratios of 1:1, 0.75:1, 0.5:1, 0.25:1, and 0.1:1. Also prepare 1% Trion (provided by CD40 kit) with RPMI 1640 medium as lysis buffer for the next step.

Tip: Preparing excess amount of each dilution group can reduce variation introduced by liquid handling and mixing.

12. Pause the RTCA, remove the E-plate, and bring to the cell culture hood. Remove 50 µL of medium from each well. The remaining volume of each well is now 100 µL.

13. Make sure the diluted NK92 cells from **step 11** are well suspended. Use a multichannel pipette to transfer 100 µL of diluted NK92 cells to the Raji cells in the wells at different E/T ratios. In "target cells only" control wells (with Raji cells only), add 100 µL of medium without NK92 cells. Add 100 µL 1% lysis buffer to wells containing Raji cells only to serve as positive controls for "fully lysed cells." Add diluted NK92 cells to the wells with medium only well (without Raji cells) as "effector cells only" controls. These wells with NK92 cells only serve as effector cell impedance background control, which is helpful for analysis.

14. Cover the lid of the E-plate and bring the E-plate back to the RTCA for continued reading.

15. Analyze the data with the RTCA software according to the software manufacturer's instructions. The instrument reading can be paused at any time to retrieve the E-plate and small samples of each treatment conditions can be removed for other orthogonal assays, for example, measuring cytokine production by ELISA or flow cytometry.

Tips: Users of the RTCA are recommended to follow the manufacturer's recommended conditions which are optimized for specific assays and analyses. One key step to ensure consistent performance of the RTCA system is to ensure target cells are seeded uniformly over the sensor on the bottom of the wells. Some cells are relatively tiny and thus may need longer than 30 min to settle down on the bottom of the well. To be certain, check target cells under the microscope to make sure they are indeed spread evenly and sitting at the bottom of the well. In our experience, after 3–4 h of seeding time, if the majority of the target cells are found sitting at or near the edge of the well, attempting to "rescue" the cells by pipetting up and down does not help with uneven cell distribution. Simply discard the E plate and repeat the cell seeding step using a new plate.

Figure 2 was generated using data exported from the RTCA software to Excel and it shows the kinetics of NK92-mediated killing of Raji cells. The treatment of Triton 0.5%, which is the final concentration in the medium of each well, serves as "full lysis" positive control.

The above protocol could be modified for same-day cell seeding and treatment.

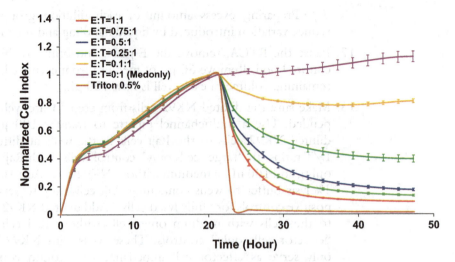

Fig. 2 Kinetics of NK92-mediated killing of Raji. RTCA software was used to display normalized Cell Index of Raji cells after the addition of NK92 cells over the indicated time course. Once 0.5% Triton-containing medium (no NK92 cells) was applied to the Raji cells, the Cell Index dropped to the minimum (dark brown line). This serves as positive control for full target cell lysis, while the negative control containing Raji cells only (medium/Med only) is represented by the pink line. All other treatment groups display target cell lysis to varying degrees in an E:T ratio-dependent manner

In **step 5**, dilute Raji cells at a density of 1,200,000 cells/mL.

In **step 6**, seed 50 μL of Raji cell suspension to bring the total volume in each well to 100 μL.

In **step 12**, skip the medium removal.

Other hematological tumor cell lines (Daudi, Ramos, Rs4;11, NALM-6, THP-1, HEL 92.1.7, MEC2, K562, MM1R, and PRMI 8226) have been tested with this protocol, but with different tethering antibodies which are optimized for the particular cell lines. For solid tumors, the RTCA can readily measure any cells attached to the sensor, which makes the assay much straightforward without the tethering procedure. An example of solid tumor killing assay using the RTCA is provided in the next protocol.

4 Assessing Cytolytic Potency of CAR-T Cells

The following protocol is designed to assess the cytolytic potency of chimeric antigen receptor (CAR)-T cells. It is important to first assess the transduction efficiency of the CAR constructs in T cells through flow-based detection of the introduced chimeric receptor on the transduced T cells.

Second generation CAR-T cells, EGFR-CD28-CAR-T cells [49, 50], are used in this protocol.

4.1 Verifying Transduction Efficiency of CAR Constructs

1. Transfer about 3×10^5 CAR-T cells and nontransduced T cells into two separate 5 mL polystyrene FACS tubes.

2. Centrifuge the tubes at $300 \times g$ for 2 min and discard the supernatant and resuspend the cells in 200 μL of FACS buffer containing 1% human serum. Divide them into 100 μL of cell suspension in two 5-mL polystyrene FACS tubes and keep all four tubes on ice for 5 min.

3. Add 1 μL of biotinylated F(ab')$_2$ fragments of goat anti-mouse F(ab')$_2$ to one tube each of CAR-T cells and nontransduced T cells. Then add 2 μL of PE-labeled anti-tag antibody and to other two tubes. Mix well and incubate all four tubes for 30 min on ice.

4. Add 3 mL of FACS buffer into each tube centrifuge at $300 \times g$ for 5 min. Discard the supernatants quickly and then vortex very briefly or shake the tubes briefly to resuspend the cell pellets in the residual liquid.

5. Add 2 μL each of APC anti-CD3 and 7-AAD antibody solution in each tube. In the tube with cells stained with anti-F(ab')$_2$ Ab, add 1 μL of PE-labeled streptavidin. Mix briefly and incubate the tubes on ice for 30 min.

6. Use FACS buffer to wash the cells again as described in **step 4** and add additional 200 μL of FACS buffer to each tube.

7. Evaluate CAR transduction efficiency by performing flow cytometry by gating on the T cells first in a forward scatter vs. side scatter plot, followed by the live cells (7-AAD-negative) in a CD3 vs. 7-AAD plot. Finally, analyze anti-F(ab')$_2$ vs. CD3 using flow software.

8. Store the EGFR-CD28-CAR-T cells in liquid nitrogen with untransduced T cells. Apply the same procedures for Mock-transduced CAR-T cells but using anti-FAB during at **step 3**.

4.2 Measuring Cytolytic Potency of EGFR-CD28 CAR-T Cells

1. Add 50 μL RPMI 1640 culture medium to each well of the E-plate 96 to determine the medium background in the RTCA as described above.

2. Use standard protocol to trypsinize BxPC3 cells from the culture flask. Add about 10 mL RPMI 1640 medium to stop the trypsinization and transfer cells to a 15 mL centrifuge tube and add fresh culture medium up to 15 mL. Pellet the cells by centrifugation for 5 min at $200 \times g$. Discard the medium and gently resuspend the cell pellet in fresh RPMI 1640 medium with a serological pipette. Count number of live BxPC3 cells.

3. Adjust the cell density to 100,000 cells/mL and transfer 100 μL of the cell suspension to each well of the E-plate.

4. Equilibrate the E-plate with BxPC3 cells at room temperature for about 30 min to allow the cells to settle evenly on the bottom of the well. This is a critical step.

5. Bring the E-plate back to the RTCA and start measuring impedance. Choose software to display Cell Index versus Time automatically at every 15 min.

6. Prepare effector EGFR-CD28-CAR-T cells at proper the desired E:T ratios in 100 μL medium (i.e., 2,000,000 cells/mL for E:T 20:1) on the following day. Include the appropriate control cells, for example, Mock or nontransduced T cells and/or unrelated CAR-T cells.

7. Pause the automatic acquisition of the RTCA and bring the E-plate from the incubator to cell culture hood.

8. Remove 50 μL of medium from each well so that the remaining medium volume is 100 μL.

9. Add 100 μL of serially diluted effector CAR-T cells or control cells at desired E:T ratios.

10. Equilibrate the E-plate at room temperature for about 30 min and return the E-plate to the RTCA to resume recording the impedance.

The RTCA reading can be paused at any time at this stage to retrieve the E-plate and small samples of each treatment conditions can be removed for other orthogonal assays, for example, measuring cytokine production by ELISA or flow cytometry.

Figure 3 displays an example of EGFR-CD28-CAR-T mediated BxPC3 cell killing over a time course. It is worth noting the background BxPC3 killing, represented by Mock and untransduced CAR-T cells, as expected [51], which could

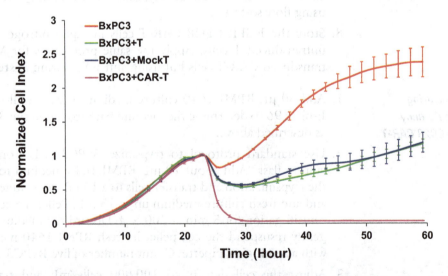

Fig. 3 Kinetics of EGFR-CD28-CAR-T cells-mediated killing of BxPC3 cancer cells BxPC3 cells are effectively lysed by EGFR-CD28-CAR-T cells (BxPC3 + CAR-T; pink line), Control groups of BxPC3 cells incubated with nontransduced T cells (BxPC3 + T; green line) or Mock-transduced CAR-T cells (BxPC3 + Mock-T) display baseline killing of target cells

be attributed to an allogeneic response since the T cells were obtained from different individuals or partial T cell activation during the manufacturing process of the CAR-T cells. The increasing Cell Index values of these two control groups over time also indicate the control T cells do not kill the target cells effectively.

Tip: The RTCA software can be used to generate other parameters for each treatment group, including effector cell-specific killing by selecting different analysis function in the RTCA Pro software.

5 Assessing BiTE-Mediated Cytolytic Potency

Bispecific T cell engaging antibody or BiTE is another clinically validated approach in immuno-oncology. BLINCYTO™ (Blinatumomab) is BiTE targeting CD19 and CD3 which has been approved by FDA for the treatment of relapsed or refractory B-cell precursor acute lymphoblastic leukemia. In this protocol, we use the RTCA to evaluate the cytolytic potency of an anti-CD3xCD19 BiTE against CD19+ B cell lymphoma cells.

5.1 Preparation of the E-Plate (Day 1)

1. Coat E-plate 96 View with 50 μL anti-CD40 antibody at the final concentration of 4 μg/mL as described above using the B cell killing kit (CD40).
2. Incubate E-plate in room temperature for 3 h.
3. Wash E-plate twice with 200 μL PBS per well.
4. Add 50 μL of RPMI 1640 medium to each well and read E-plate in the RTCA to determine background as described above.

5.2 Seeding of Daudi Cells (Day 1)

1. Count and transfer Daudi cells (50,000 cells/well) to a centrifuge tube.
2. Centrifuge cells at 300 × *g* for 5 min at RT and discard the supernatant.
3. Resuspend cells in RPMI 1640 full medium at the density of 500,000 cells/mL, mix gently and well.
4. Seed Daudi cells onto E-plate at 50,000 cells/well in 100 μL volume.
5. Leave the E-plate with Daudi cells in the cell culture hood for 30 min.
6. Monitor the Daudi cells in the RTCA overnight.

5.3 Preparation of the Effector Cells and Treatment (Day 2)

We recommend using freshly prepared peripheral blood mononuclear cells (PBMCs) from whole peripheral blood by density gradient centrifugation. This should be carried out inside the cell culture hood for sterilization purpose.

1. Prepare fresh PBMCs from healthy donors using Ficoll-Paque Plus (GE Healthcare) according to the manufacturer's protocol. Briefly, freshly collected blood is diluted with twofold volume of PBS (Ca/Mg, −/−). The diluted blood is then gently loaded on the top of 15 mL Ficoll-Paque Plus solution in a 50 mL Falcon tube. Centrifuge the tube at 400 × g for 30 min at room temperature and stop the centrifugation with break off. The supernatant is removed and PBMCs are then collected from the concentrated layer with PBS (Ca/Mg, −/−) for another spin in 400 × g. The remaining cells are then washed two to three times with PBS (Ca/Mg, −/−). Finally, the PBMCs are collected and resuspended in RPMI 1640 with 10% FBS and Pen/Strep. Once PBMCs are ready, they are either cultured briefly in the above medium or immediately subjected to the T cell isolation.

   TIP: If frozen PBMC are used, resuspend cells in RPMI 1640 full medium at about 2,500,000 cells/mL. Centrifuge the cells at 350 × g for 5 min at RT and discard the supernatant. Culture the cells in fresh RPMI 1640 full medium for 2–4 h.

2. Resuspend PBMC cell suspension at a concentration of 50,000,000 cells/mL in recommended buffer (PBS + 2% FBS + 1 mM EDTA, Ca^{2+} and Mg^{2+} free). Transfer cells to a 5 mL (12 × 75 mm) polystyrene tube.

3. Add the EasySep™ Human T Cell Enrichment Cocktail at 50 μL/mL cells (e.g., for 2 mL of cells, add 100 μL of cocktail). Mix well and incubate at room temperature (15–25 °C) for 10 min.

4. Vortex the EasySep™ D Magnetic Particles for 30 s. Ensure that the particles are in a uniform suspension with no visible aggregates.

5. Add the EasySep™ D Magnetic Particles at 50 μL/mL cells (e.g., for 2 mL of cells, add 100 μL of magnetic particles). Mix well and incubate at room temperature (15–25 °C) for 5 min.

6. Bring the cell suspension up to a total volume of 2.5 mL by adding PBS free of Ca/Mg with 2% FBS and 1 mM EDTA. Mix the cells in the tube by gently pipetting up and down for two to three times. Place the tube (without cap) into the magnet. Set aside for 5 min.

7. Pick up the EasySep™ Magnet, and in one continuous motion invert the magnet and tube, pouring off the desired fraction into a new 5 mL polystyrene tube.

8. Centrifuge T cells at 300 × g for 5 min at RT and discard the supernatant. Resuspend T cells in RPMI 1640 full medium at 5,000,000 cells/mL.

9. Pause the RTCA and bring the E-plate to cell culture hood to add T cells to the treatments along with other controls based on the experiment design.

10. Add 1,000,000 T cells in 100 μL to Daudi cells in the well at the ratio of E:T = 20:1.

11. Add BiTE antibody (330 ng/mL) and same amount of anti-CD19 antibody in the control wells.

12. Return the E-plate back to the RTCA to resume monitoring. The software can display the Cell index or % cytolysis versus time in real time.

Once the treatment started, the RTCA system could be paused at any time and part of the treatment cells or solution could be sampled and subjected to other orthogonal assays, that is, end point analysis by flow cytometry. After 48 h of the treatment, samples from different wells are obtained and subjected to 7-AAD based flow cytometer analysis. The normalized % cytolysis measured by FLOW and RTCA for Daudi + T + BiTE are 74.02 ± 1.12% and 72.88 ± 5.11%, respectively. An example of the RTCA result is shown in Fig. 4.

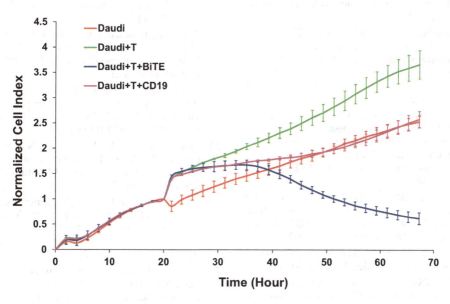

Fig. 4 Anti-CD19xCD3 BiTE mediated T cell killing of the Daudi cells. Daudi cells were seeded on anti-CD40 coated plate for about 20 h. Purified T cells were added with anti-CD19 antibody (300 ng/mL) as control. Treatment of Daudi with T cells plus anti-CD19xCD3 BiTE (300 ng/mL) exhibited specific Daudi killing over control groups (Daudi + T cells only and Daudi + T cells + anti-CD19 antibody). A similar result was observed if PBMCs were applied to the assay instead of purified T cells

References

1. Williams O (2004) Flow cytometry-based methods for apoptosis detection in lymphoid cells. Methods Mol Biol 282:31–42. https://doi.org/10.1385/1-59259-812-9:031
2. Wlodkowic D, Skommer J, Darzynkiewicz Z (2009) Flow cytometry-based apoptosis detection. Methods Mol Biol 559:19–32. https://doi.org/10.1007/978-1-60327-017-5_2
3. Suzuki T, Fujikura K, Higashiyama T, Takata K (1997) DNA staining for fluorescence and laser confocal microscopy. J Histochem Cytochem 45(1):49–53. https://doi.org/10.1177/002215549704500107
4. Petty RD, Sutherland LA, Hunter EM, Cree IA (1995) Comparison of MTT and ATP-based assays for the measurement of viable cell number. J Biolumin Chemilumin 10(1):29–34. https://doi.org/10.1002/bio.1170100105
5. Crouch SPKR, Slater KJ, Fletcher J (1993) The use of ATP bioluminescence as a measure of cell proliferation and cytotoxicity. J Immunol Methods 160(1):81–88
6. Angioli R, Sevin BU, Perras JP, Untch M, Koechli OR, Nguyen HN, Steren A, Schwade JG, Villani C, Averette HE (1993) In vitro potentiation of radiation cytotoxicity by recombinant interferons in cervical cancer cell lines. Cancer 71(11):3717–3725
7. Holden HT, Oldham RK, Ortaldo JR, Herberman RB (1977) Standardization of the chromium-51 release, cell-mediated cytotoxicity assay: cryopreservation of mouse effector and target cells. J Natl Cancer Inst 58(3):611–622
8. Nelson DL, Kurman CC, Serbousek DE (2001) 51Cr release assay of antibody-dependent cell-mediated cytotoxicity (ADCC). Curr Protoc Immunol 7:27. https://doi.org/10.1002/0471142735.im0727s08
9. Hamburger AW (1987) The human tumor clonogenic assay as a model system in cell biology. Int J Cell Cloning 5(2):89–107. https://doi.org/10.1002/stem.5530050202
10. Soderdahl DW (1988) The clonogenic assay in perspective. Hawaii Med J 47(4):151–152
11. EA T (1989) Human tumor clonogenic assay: what is new? Eur J Cancer Clin Oncol 25(7):1031–1033
12. Pittet MJSD, Valmori D, Rimoldi D, Liénard D, Lejeune F, Cerottini JC, Romero P (2001) Ex vivo analysis of tumor antigen specific CD8+ T cell responses using MHC/peptide tetramers in cancer patients. Int Immunopharmacol 1(7):1235–1247
13. Bercovici N, Duffour MT, Agrawal S, Salcedo M, Abastado JP (2000) New methods for assessing T-cell responses. Clin Diagn Lab Immunol 7(6):859–864
14. Shacklett BL (2002) Beyond 51Cr release: new methods for assessing HIV-1-specific CD8+ T cell responses in peripheral blood and mucosal tissues. Clin Exp Immunol 130(2):172–182
15. Sitz KV, Birx DL (1999) Lymphocyte proliferation assay. Methods Mol Med 17:343–353. https://doi.org/10.1385/0-89603-369-4:343
16. Mond JJ, Brunswick M (2003) Proliferative assays for B cell function. Curr Protoc Immunol 3:10. https://doi.org/10.1002/0471142735.im0310s57
17. Kruisbeek AM, Shevach E, Thornton AM (2004) Proliferative assays for T cell function. Curr Protoc Immunol 3:12. https://doi.org/10.1002/0471142735.im0312s60
18. Malyguine AM, Strobl S, Dunham K, Shurin MR, Sayers TJ (2012) ELISPOT assay for monitoring cytotoxic T lymphocytes (CTL) activity in cancer vaccine clinical trials. Cell 1(2):111–126. https://doi.org/10.3390/cells1020111
19. Ranieri E, Popescu I, Gigante M (2014) CTL ELISPOT assay. Methods Mol Biol 1186:75–86. https://doi.org/10.1007/978-1-4939-1158-5_6
20. Qiu JG, Mei XL, Chen ZS, Shi Z (2014) Cytokine detection by flow cytometry. Methods Mol Biol 1172:235–242. https://doi.org/10.1007/978-1-4939-0928-5_21
21. Jung T, Schauer U, Heusser C, Neumann C, Rieger C (1993) Detection of intracellular cytokines by flow cytometry. J Immunol Methods 159(1–2):197–207
22. Foster B, Prussin C, Liu F, Whitmire JK, Whitton JL (2007) Detection of intracellular cytokines by flow cytometry. Curr Protoc Immunol 6:24. https://doi.org/10.1002/0471142735.im0624s78
23. De Haan A, Van Der Gun I, Van Der Bij W, De Leij LF, Prop J (2002) Detection of alloreactive T cells by flow cytometry: a new test compared with limiting dilution assay. Transplantation 74(4):562–570
24. Ramis G, Martinez-Alarcon L, Quereda JJ, Mendonca L, Majado MJ, Gomez-Coelho K, Mrowiec A, Herrero-Medrano JM, Abellaneda JM, Pallares FJ, Rios A, Ramirez P, Munoz A

(2013) Optimization of cytotoxicity assay by real-time, impedance-based cell analysis. Biomed Microdevices 15(6):985–995. https://doi.org/10.1007/s10544-013-9790-8
25. Fasbender F, Watzl C (2018) Impedance-based analysis of natural killer cell stimulation. Sci Rep 8(1):4938. https://doi.org/10.1038/s41598-018-23368-5
26. Cerignoli F, Abassi YA, Lamarche BJ, Guenther G, Ana DS, Guimet D, Zhang W, Zhang J, Xi B (2018) In vitro immunotherapy potency assays using real-time cell analysis. PLoS One 13:e0193498. https://doi.org/10.1371/journal.pone.0193498
27. Halle S, Halle O, Forster R (2017) Mechanisms and dynamics of T cell-mediated cytotoxicity in vivo. Trends Immunol 38(6):432–443. https://doi.org/10.1016/j.it.2017.04.002
28. Somanchi SS, McCulley KJ, Somanchi A, Chan LL, Lee DA (2015) A novel method for assessment of natural killer cell cytotoxicity using image cytometry. PLoS One 10(10):e0141074
29. Peper JK, Schuster H, Loffler MW, Schmid-Horch B, Rammensee HG, Stevanovic S (2014) An impedance-based cytotoxicity assay for real-time and label-free assessment of T-cell-mediated killing of adherent cells. J Immunol Methods 405:192–198. https://doi.org/10.1016/j.jim.2014.01.012
30. Ke N, Wang X, Xu X, Abassi YA (2011) The xCELLigence system for real-time and label-free monitoring of cell viability. Methods Mol Biol 740:33–43. https://doi.org/10.1007/978-1-61779-108-6_6
31. McGuinness R (2007) Impedance-based cellular assay technologies: recent advances, future promise. Curr Opin Pharmacol 7(5):535–540. https://doi.org/10.1016/j.coph.2007.08.004
32. Xi B, Yu N, Wang X, Xu X, Abassi YA (2008) The application of cell-based label-free technology in drug discovery. Biotechnol J 3(4):484–495. https://doi.org/10.1002/biot.200800020
33. Atienza JM, Yu N, Kirstein SL, Xi B, Wang X, Xu X, Abassi YA (2006) Dynamic and label-free cell-based assays using the real-time cell electronic sensing system. Assay Drug Dev Technol 4(5):597–607
34. Giaever I, Keese CR (1984) Monitoring fibroblast behavior in tissue culture with an applied electric field. Proc Natl Acad Sci U S A 81(12):3761–3764
35. Scott CW, Peters MF (2010) Label-free whole-cell assays: expanding the scope of GPCR screening. Drug Discov Today 15(17–18):704–716. https://doi.org/10.1016/j.drudis.2010.06.008
36. Lundstrom K (2017) Cell-impedance-based label-free technology for the identification of new drugs. Expert Opin Drug Discov 12(4):335–343. https://doi.org/10.1080/17460441.2017.1297419
37. Atienza JM, Yu N, Wang X, Xu X, Abassi Y (2006) Label-free and real-time cell-based kinase assay for screening selective and potent receptor tyrosine kinase inhibitors using microelectronic sensor array. J Biomol Screen 11(6):634–643
38. Xi B, Wang T, Li N, Ouyang W, Zhang W, Wu J, Xu X, Wang X, Abassi YA (2011) Functional cardiotoxicity profiling and screening using the xCELLigence RTCA cardio system. J Lab Autom 16(6):415–421
39. Abassi YA, Xi B, Li N, Ouyang W, Seiler A, Watzele M, Kettenhofen R, Bohlen H, Ehlich A, Kolossov E, Wang X, Xu X (2012) Dynamic monitoring of beating periodicity of stem cell-derived cardiomyocytes as a predictive tool for preclinical safety assessment. Br J Pharmacol 165(5):1424–1441
40. Abassi YA, Jackson JA, Zhu J, O'Connell J, Wang X, Xu X (2004) Label-free, real-time monitoring of IgE-mediated mast cell activation on microelectronic cell sensor arrays. J Immunol Methods 292(1–2):195–205
41. Abassi YA, Xi B, Zhang W, Ye P, Kirstein SL, Gaylord MR, Feinstein SC, Wang X, Xu X (2009) Kinetic cell-based morphological screening: prediction of mechanism of compound action and off-target effects. Chem Biol 16(7):712–723. https://doi.org/10.1016/j.chembiol.2009.05.011
42. Moodley K, Angel CE, Glass M, Graham ES (2011) Real-time profiling of NK cell killing of human astrocytes using xCELLigence technology. J Neurosci Methods 200(2):173–180
43. Hegde M, Mukherjee M, Grada Z, Pignata A, Landi D, Navai SA, Wakefield A, Fousek K, Bielamowicz K, Chow KK, Brawley VS, Byrd TT, Krebs S, Gottschalk S, Wels WS, Baker ML, Dotti G, Mamonkin M, Brenner MK, Orange JS, Ahmed N (2016) Tandem CAR T cells targeting HER2 and IL13Ralpha2 mitigate tumor antigen escape. J Clin Invest 126(8):3036–3052. https://doi.org/10.1172/JCI83416
44. Hillerdal V, Boura VF, Bjorkelund H, Andersson K, Essand M (2016) Avidity characterization of genetically engineered T-cells with novel and established approaches. BMC Immunol 17(1):23. https://doi.org/10.1186/s12865-016-0162-z

45. Jin J, Gkitsas N, Fellowes VS, Ren J, Feldman SA, Hinrichs CS, Stroncek DF, Highfill SL (2018) Enhanced clinical-scale manufacturing of TCR transduced T-cells using closed culture system modules. J Transl Med 16(1):13. https://doi.org/10.1186/s12967-018-1384-z

46. Freedman JD, Hagel J, Scott EM, Psallidas I, Gupta A, Spiers L, Miller P, Kanellakis N, Ashfield R, Fisher KD, Duffy MR, Seymour LW (2017) Oncolytic adenovirus expressing bispecific antibody targets T-cell cytotoxicity in cancer biopsies. EMBO Mol Med 9 (8):1067–1087. https://doi.org/10.15252/emmm.201707567

47. Seidel UJ, Vogt F, Grosse-Hovest L, Jung G, Handgretinger R, Lang P (2014) Gammadelta T cell-mediated antibody-dependent cellular cytotoxicity with CD19 antibodies assessed by an impedance-based label-free real-time cytotoxicity assay. Front Immunol 5:618. https://doi.org/10.3389/fimmu.2014.00618

48. Kute T, Stehle JR, Ornelles D, Walker N, Delbono O, Vaughn JP (2012) Understanding key assay parameters that affect measurements of trastuzumab-mediated ADCC against Her2 positive breast cancer cells. Oncoimmunology 1(6):810–821. https://doi.org/10.4161/onci.20447

49. Zhou Y, Drummond DC, Zou H, Hayes ME, Adams GP, Kirpotin DB, Marks JD (2007) Impact of single-chain Fv antibody fragment affinity on nanoparticle targeting of epidermal growth factor receptor-expressing tumor cells. J Mol Biol 371(4):934–947. https://doi.org/10.1016/j.jmb.2007.05.011

50. Liu X, Jiang S, Fang C, Yang S, Olalere D, Pequignot EC, Cogdill AP, Li N, Ramones M, Granda B, Zhou L, Loew A, Young RM, June CH, Zhao Y (2015) Affinity-tuned ErbB2 or EGFR chimeric antigen receptor T cells exhibit an increased therapeutic index against tumors in mice. Cancer Res 75(17):3596–3607. https://doi.org/10.1158/0008-5472.CAN-15-0159

51. Golubovskaya V, Berahovich R, Zhou H, Xu S, Harto H, Li L, Chao CC, Mao MM, Wu L (2017) CD47-CAR-T cells effectively kill target cancer cells and block pancreatic tumor growth. Cancers (Basel) 9(10):E139. https://doi.org/10.3390/cancers9100139

Chapter 4

Assessing the Potency of T Cell-Redirecting Therapeutics Using In Vitro Cancer Cell Killing Assays

Tomasz Dobrzycki, Andreea Ciuntu, Andrea Stacey, Joseph D. Dukes, and Andrew D. Whale

Abstract

The development of in vitro cell-based assays that can best recapitulate as near as possible to a human in vivo setting is urgently needed, as animal models are unsuitable for the preclinical assessment of the growing number of fully human cancer therapeutics. In the field of immune-oncology (I/O), interest is focused on efforts to better understand and manipulate the immune system to either overcome tolerance or promote tumor cell killing. However, T cell responses are frequently assessed using in vitro cell-based assays that have traditionally focused on simple endpoint measurements, such as cytokine release or cell proliferation, which do not represent a true readout of potency (i.e., the ability of T cells to induce apoptosis of cancer cells). In addition, traditional T cell killing assays, including Cr^{51} or lactate dehydrogenase (LDH) release assays, usually have short-term end points, and use reagents or technology that can present technical challenges. Furthermore, from the drug discovery perspective, these assays are not well suited to high throughput 96 or 384-well screening assay formats.

In this chapter, we describe in detail reliable methods to quantify anticancer cell activity mediated by T cell killing in extended duration and kinetic assays using real-time measurement of apoptosis or cell lysis. We offer a comparison of the relative merits of two platforms, IncuCyte (a Sartorius brand) and xCELLigence (ACEA Biosciences), highlighting critical areas of assay optimization and discuss technical considerations. We have found these methods to be robust and believe them suitable for testing a broad spectrum of I/O therapeutic modalities either alone or in combination with additional agents. Their broad applicability will also be of value to researchers undertaking basic research as well as drug discovery in I/O or other therapeutic areas associated with measuring T cell responses.

Key words Cancer, Immunotherapy, T cell, TCR, Bispecifics, T cell killing, Cell-based assays, ImmTAC

1 Introduction

The preclinical testing of new drugs has classically relied on the use of animal models to evaluate potency as a predictive readout of clinical efficacy and, in some cases, an indication of safety

---

Tomasz Dobrzycki and Andreea Ciuntu contributed equally to this work.

Seng-Lai Tan (ed.), *Immuno-Oncology: Cellular and Translational Approaches*, Methods in Pharmacology and Toxicology, https://doi.org/10.1007/978-1-0716-0171-6_4, © Springer Science+Business Media, LLC, part of Springer Nature 2020

[1]. However, the growing number of fully human-specific cancer therapeutics necessitates the development of robust in vitro cell-based assays that can best recapitulate a human in vivo setting. There are two key challenges for developing assays that can adequately test molecules capable of co-opting a patient's immune system to produce an effective anticancer response [2]. Firstly, because the mode of action of I/O therapeutics requires the ability to manipulate or modulate immune cells to kill tumor cells, a model system requires the coculture of both cancer cells and immune cells. Secondly, readout of T cell activation has traditionally focused on simple endpoint measurements, such as cytokine release or induction of cell proliferation. Although these approaches are informative, they offer little kinetic information, nor do they represent a true readout of potency (i.e., the ability of T cells to induce apoptosis of cancer cells).

Preclinical assessment of the ability of a therapy to mediate or redirect T cells to kill cancer cells provides a relevant measurement of pharmacological potency and can be indicative of efficacy. Traditional assays that measure T cell killing, Cr^{51} or LDH release, or cytotoxicity flow cytometry assays have a number of limitations. Cr^{51} and LDH release assays suffer from high background that results from spontaneous leakage of the label from the target cancer cells and limits their utility to short-term (<24 h) endpoint assays. Furthermore, these assays are not readily amenable to high throughput 96 or 384-well formats [3, 4].

An ideal assay format for the assessment of redirected T cell killing would involve the use of easy-to-handle, non-perturbing labels (or be label free) and have the capability of monitoring cells over an extended time span (>24 h) in high throughput multi-well format plates. Recently, several commercial technology platforms have been developed with these capabilities and, as a result, the potential to overcome the limitations of traditional assays. The IncuCyte® (a Sartorius brand) and ACEA Biosciences' xCELLigence system are two such platforms, developed to allow the incorporation of assay measurements within cell culture incubators to enable dissection of immune and cancer cell interactions in a live cell assay format.

The IncuCyte® system was developed by Essen Bioscience, a Sartorius company, in 2006 in its first iteration, the FLR. The current system, the S3, is the third iteration following the ZOOM; however, all models are hereafter referred to collectively as "IncuCyte." The IncuCyte platform is capable of microscopy image-based data collection from up to six microplates at a time using an automated moving objective. The system allows for the real-time, automated, time-lapse monitoring of live cells. Coupled with fluorescent probes, embedded software can further distinguish between live and dead cells in captured images [5].

The ACEA Biosciences' xCELLigence Real-Time Cell Analyzer (RTCA) device consists of four components: RTCA Analyzer, RTCA Station, RTCA Control Unit, and a specialized E-Plate 96-well plate that uses microelectronic biosensors to monitor electrical impedance within cellular assays (see Chap. 3). The RTCA MP instrument allows simultaneous, yet independent, analysis of up to six 96-well plates [6]. Electrical impedance, which is affected by the presence of cells, is measured by the RTCA Analyzer in each well of an E-Plate 96. The cells alter the local ionic environment, increasing impedance at the electrodes; as such, impedance depends on and is a measure of the strength of cell adhesion and the number of cells present within a well [6].

In this chapter, we describe detailed methodology for the use of the IncuCyte and xCELLigence platforms to measure T cell killing of target cells in real time. The main benefit of both platforms is the ability to conduct longer time span, kinetic assays with real time measurement of anticancer cell activity mediated by T cell killing. Data is derived from either direct visualization of T cell killing or quantification of target cell lysis. Both can be invaluable for either basic research or the evaluation of molecules designed to activate T cell killing during drug discovery. We offer a comparison of the relative merits of the two platforms (*see* **Note 1**), highlighting key aspects of assay optimization, and discuss various technical considerations. These insights are based on over 10 years' experience of practical use working with cocultures of primary human T cells, cancer cell lines, and soluble T cell engagers (sTE) [7, 8]. sTE are bifunctional molecules that bind both tumor antigens and the T cell receptor (TCR) complex. sTE redirect or target T cells to the tumor via recruitment and activation of T cells irrespective of their natural specificity [9]. Some sTEs are based on monoclonal antibodies, such as *bi*specific T cell engagers (BiTE®),while ImmTAC® (*Immune mobilising monoclonal TCR Against Cancer*) molecules are soluble picomolar affinity-engineered TCRs that recognize tumor antigen by engaging peptides presented in the context of human leukocyte antigen (HLA) on the surface of cancer cells, fused to an anti-CD3 scFv domain [10]. ImmTAC molecules potently and specifically redirect T cells, mediating tumor cell killing in vitro and in vivo [11]. Moreover, tebentafusp, an ImmTAC specific for a gp100 peptide presented in the context of HLA-A*02, is showing signs of clinical efficacy in solid tumors [12]. Although our experience has focused on assessment of the potency of TCR-based ImmTAC molecules, we believe the methods presented here can be easily modified for testing a broad spectrum of I/O therapeutic modalities alone or in combination with additional agents, as well as alternative target cells appropriate to indications other than oncology. Finally, we suggest additional opportunities to augment the data by multiplexing redirected T cell killing assays with conventional endpoint measurement assays to quantify markers of T cell activation, such as cell proliferation or cytokine release.

2 Materials

2.1 Common Reagents

1. Ca^{2+} and Mg^{2+} free Dulbecco's PBS, pH 7.4 (Gibco).
2. 0.25% Trypsin-EDTA (Gibco).
3. ViaStain™AOPI staining solution (Nexcelom Bioscience).
4. Assay media: RPMI 1640 buffered with 25 mM HEPES (Gibco) + 10% heat-inactivated fetal bovine serum (FBS) supplemented with 0.2 mM L-glutamine and 50 U/mL penicillin and 50 μg/mL streptomycin.
5. 30 μm pre-separation filter (MiltenyiBiotec).
6. Soluble T cell redirecting test reagent for example, ImmTAC molecule.
7. Effector cells (*see* **Note 2**), for example, primary freshly isolated or cryopreserved CD8+ T cells or peripheral blood mononuclear cells (PBMC).
8. Target cells (*see* **Notes 3–5**).

2.2 IncuCyte-Specific Reagents

1. 96-well flat bottom tissue culture plate with lid. We routinely use standard sterile tissue culture treated clear flat-bottomed plates or ImageLock plates (Sartorius; these plates are etched to provide a fiducial marker that we have found beneficial, especially when imaging in two fluorescent channels (*see* **Note 4**) or multiplexing with endpoint assays to measure cytokine release (*see* **Note 7**)).
2. IncuCyte Caspase-3/7 Green Apoptosis Assay Reagent (activated caspase-3/7 sensitive recognition motif (DEVD) coupled to NucView™488) (Sartorius).
3. IncuCyte ZOOM or S3 (Sartorius).

2.3 xCELLigence-Specific Reagents

1. xCELLigence E-plate 96 PET (ACEA Biosciences).
2. xCELLigence RTCA MP plate reader (ACEA Biosciences).

2.4 Reagents for Multiplexing

1. Cell Trace™ Violet (CTV) Cell Proliferation Kit (Life Technologies).
2. FACS buffer: 1× PBS supplemented with 2% FBS and 2 mM EDTA.
3. Sample fixative, such as Cytofix™ (BecktonDickinson) or 2% paraformaldehyde (PFA) in cold, filtered 1× PHEM buffer (2× PHEM buffer: 14 g PIPES, 6.5 g HEPES, 3.8 g EGTA, 0.99 g MgSO4 (pH to 7.2).

4. Antibodies (*see* **Note 7**):

   Anti-CD8 APC clone RPA-T8 and anti-CD4 PE-Cy7 clone RPA-T4 (eBioscience).

   Anti-CD3 APC-Cy7 clone HIT3a (Biolegend).

5. Zombie Green™ live/dead stain (Biolegend).

3 Methods

3.1 Principle of Assay Design

We have found the methods presented here (which are predominantly based on the use of immortalized cancer cell lines and cryopreserved PBMC) to be robust for the assessment of potency of TCR-based bispecific T cell redirecting ImmTAC molecules. However, because of the inherent variability associated with primary immune cells, it may be necessary to optimize some of the key parameters of the assay, including the ratio of effector cells to target cells (E:T) and time course of the assay. We advocate an assay time frame of 72–96 h because coculturing cells for longer periods without a change of media may result in the depletion of nutrients, which could affect T cell function or general cell viability, thereby complicating the interpretation of results. We recommend using standard assay development practices based on empirical determination and titration.

As a general rule, assay plates should be set up with biological triplicates for all conditions tested. In addition to evaluating the effect of a dose response of test T cell redirecting reagent on cocultures of target and effector cells, we have found the following controls to be critical to evaluate both the assay performance and post-hoc analysis:

1. PBMC or purified T cell effectors (alone and with test reagent).

2. Target (antigen expressing) cancer cells (alone and with test reagent).

3. No drug controls, that is, cocultures of targets and effectors.

4. Negative control target cell lines (that either naturally do not express antigen or have been engineered or modified to remove antigen expression).

5. Positive controls: These may include agents to induce apoptosis, for example, a cocktail of cytotoxic drugs or chemicals (e.g., 0.4% Triton X-100) to mediate target cell lysis or more-specific controls such as a previously characterized sTE molecule or anti-CD3/28 antibodies to activate T cells. For ImmTAC molecules, pulsing target cells with an ImmTAC-specific peptide can be used to induce redirected T cell killing.

Irrespective of platform or cell type, an essential assay parameter to optimize is the seeding density of target cells. In our experience, an ideal target cell seeding density is one that results in approximately 20% confluence at assay initiation to allow the cell line to maintain consistent, near logarithmic growth for a 72-h period (the duration of the assay). If cells reach confluence before 72 h, they may begin to self-limit proliferation, for example, by induction of apoptosis, detach from the assay plate and/or deplete assay media of essential components, all of which have the potential to confound the analysis by contributing to a positive signal or by negatively impacting on T cell function. For the xCELLigence assay, an additional consideration is to select a seeding density that results in a calculated Cell Index (CI) of 1 or as close as possible to 1, at 24 h (*see* Subheadings 3.2 and 4.2). Optimal seeding densities will largely depend on the size and doubling time of the target cell (line) and are best determined empirically.

3.2 Protocol for Optimizing Seeding Density of Target Cells

Plate target cells (in triplicate) in an appropriate assay plate (refer to protocol Subheadings 3.3 and 3.4) to yield an appropriate range of cell numbers per well. The optimal seeding density for most cell lines usually falls within the range of 3000–30,000 cells per well. Therefore, an ideal range to evaluate would include seeding densities that are lower and greater than the anticipated optimum, for example, 40,000, 20,000, 10,000, 5000, 2500, 1250, and 625. The optimal number of cells per well and range tested for specific cell lines will vary depending on cell size, doubling time, and kinetics of cell proliferation.

To determine optimal cell seeding density using the IncuCyte:

1. Seed cells in 100 μL media and leave plate at room temperature for 30 min before incubating overnight at 37 °C to allow sufficient time for the cells to adhere and spread to the bottom of the well. We have found the inclusion of a short room temperature equilibration step helps promote even distribution of cells to assays plates.

2. Add an additional 100 μL media to each well of the assay plate and place plate in the microplate tray and load into the IncuCyte equipped with a 10× objective.

3. Incubate the plates for 30 min to allow condensation to evaporate before setting to acquire images in bright field at 2- to 3-h intervals for 96 h.

4. At the end of the assay, the rate of cell proliferation and density can be analyzed using the confluence metric following manufacturer's recommendations [5] to identify an optimal seeding density (Fig. 1a).

To determine optimal cell seeding density using the xCELLigence:

1. Set up RTCA Station as described in Subheading 3.4, pipette 50 μL of assay media into each well of the E-plate and measure the background impedance of the assay medium.

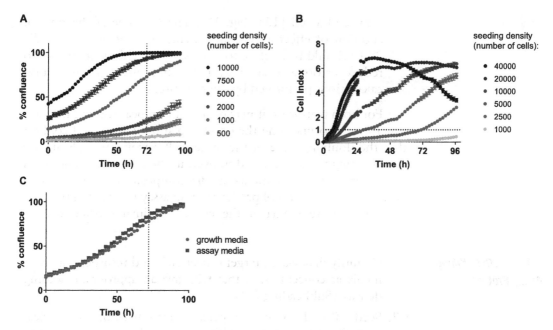

Fig. 1 Seeding density optimization. (a) Pancreatic cancer cells were plated at a range of seeding densities indicated by starting number of cells per well, and time-dependent cell proliferation was monitored using an IncuCyte ZOOM kinetic live cell imaging system to image and quantify confluence at 2-h intervals for a 96-h period. The optimal seeding density is indicated in blue. (b) Pancreatic cancer cells were plated at indicated starting number of cells per well, and cell proliferation was monitored using an RTCA xCELLigence plate reader to quantify electrical impedance at 2-h intervals for a 96-h period. The optimal seeding density is indicated in blue. Note that high starting numbers result in rapid saturation of impedance and subsequent drop in CI, associated with apoptosis of over-confluent cells. The indicated optimal starting number of cells allowed a steady increase in CI throughout the assay, indicative of exponential growth of the cells, and reached a CI value close to 1 at 24 h. (c) Example of an experiment to compare the effect of cell proliferation in assay and standard culture (growth) media for a melanoma cell line. Confluence was monitored using an IncuCyte ZOOM kinetic live cell imaging system to image and quantify confluence at 2-h intervals for a 96-h period. Error bars indicate S.E.M. of replicate samples

2. Seed target cells at a range of densities in 100 μL assay media (total volume 150 μL).

3. Leave plate to equilibrate for 30 min at room temperature.

4. Return the plate to the RTCA Station and set to perform sweeps at 30-min intervals.

5. After 24 h, remove plate from RTCA Station and remove 100 μL of assay medium and replace with 150 μL of pre-warmed medium. This step mimics the protocol used in the T cell-mediated killing assay.

6. Return plate to the RTCA Station and set to record CI measurements every 2 h for an additional 72–96 h.

7. At the end of the assay, the rate of cell proliferation and optimal cell density can be analyzed using the measured impedance,

expressed as CI [13] (Fig. 1b). The CI value at the time of addition of effector cells and therapeutic molecules (usually 24 h) should be as close to 1 as possible. This is important to reduce errors associated with the subsequent normalization calculations performed by the software.

For both platforms, it may be useful to repeat seeding titration experiments to fine-tune the optimal plating density and confirm that the doubling time of the target cell remains broadly similar across a range of passages. Moreover, in the case of cell lines with specific growth requirements/media composition, it is useful to perform a comparison of proliferation in assay media with proliferation in culture media at the chosen optimal seeding density (Fig. 1c).

3.3 IncuCyte Killing Assay Protocol

1. Quantify dissociated target cancer cells and resuspend in assay media at concentration that will give an appropriate seeding density (Subheading 3.2).

2. Seed 100 µL of cell suspension into flat-bottomed, clear 96-well plate compatible with use on the IncuCyte, equilibrate at room temperature for 30 min, and incubate at 37 °C overnight.
   NB: The IncuCyte is calibrated to use plates from all major manufacturers and suppliers.

3. Aspirate media and replace with 50 µL assay media.

4. Prepare the following reagents, diluting stocks into assay media. Typically, we generate each component as a 3 or 6× concentrated working stock, and seed each at 25 or 50 µL for a total of 150 µL per well in the assay plate.

   (a) Fluorescent Caspase-3/7 sensitive apoptotic marker reagent. Dilute as appropriate to achieve a 5 µm final concentration in the well. Alternative Annex in V reagents are available (Sartorius) and should be made up and used in accordance with supplier's instructions.

   (b) Soluble test agent, for example, T cell redirecting ImmTAC or BiTE(s).

   (c) PBMC (refer to **Note 2**). Preparation of PBMC should be left until the final step to ensure maximum viability and function of the effectors.

5. Transfer each component into the assay plate(s) sequentially using reverse pipetting to minimize or prevent the introduction of air bubbles that can impair image acquisition at early time points. Adjust volume in control wells that lack individual components using assay media.

6. Place assay plates in the IncuCyte and wait a minimum of 30 min to allow condensation to dissipate before starting

image acquisition. Set the IncuCyte, equipped with a 10× objective, to acquire images in bright field and the fluorescence channel appropriate for detection of apoptotic marker every 2 h for a ~96-h time frame.

7. Image analysis should be performed according to the manufacturer's instructions [5, 14]. Briefly, experimental analysis is performed in four stages: generation of an image collection, creation of a processing definition to analyze each image, running the analysis, and quality control (QC) of the analysis. In our experience, the points discussed below are useful considerations and starting points.

8. Add images to a new image collection by clicking on Create or Add to Image Collection link. The Image Collection is used to adjust the parameters for the analysis. The following points should be taken into account when selecting images:

    (a) The speed of analysis is greatly improved by limiting the size of the image collection. Three or four well-chosen and representative images are sufficient to train and refine the analysis.

    (b) Ideally, an image collection should consist of images captured from wells in which killing can be clearly visualized at both an early (<24 h) and late (>48 h) time point, as well as negative control wells containing target cells alone and effector cells alone. Image collections for assays performed with targets labeled in a second channel should also be included as additional controls (*see* **Note 4**).

9. Generate a processing definition based on the newly created image collection. The processing definition can be adjusted to subtract background fluorescence and apply an array of filters. The goal is to write a process definition capable of detecting target cells that are fluorescent due to uptake of the apoptotic marker as a result of redirected T cell killing. Because apoptotic T cells will also fluoresce, accurate analysis and interpretation of IncuCyte killing assay data require a process definition that effectively discriminates between target and effector populations. In our experience, this is best achieved by dispensing with analysis of the confluence measured from bright field images in favor of focusing on the fluorescence channel used to detect apoptotic marker. We have found the following parameters to be useful starting points:

    (a) In accordance with manufacturer's recommendations, set the background subtraction method to "Top-HAT."

    (b) Adjust the radius filter in the green channels to 100 μm and minimum area to 125 μm^2. These filters can be adjusted to discriminate targets from effectors (cancer cell targets are usually much bigger than effector PBMC or T cells). Clicking on preview applies the filters to the image. By hovering the cursor over an object in an image that is within the filter or is considered background, it is possible to see which factors have resulted in the inclusion or exclusion of the object and adjust the filters if necessary.

(c) For target cell lines that cluster together or grow in colonies and to account for activated T cells that may cluster or blast [15], select Edge Split ON and use the slide bar to adjust the sensitivity to better define the objects in a cluster. Click on Preview to apply, view, and, if necessary, adjust the filter.

(d) The eccentricity mean intensity and integrated intensity parameters can also be adjusted to remove background, dead cells, and debris. However, most of the processing definition should have been adjusted by applying the parameters described in **steps a–c**.

10. Run the analysis. To QC the process definition, interrogate the control wells containing target cells only, effector cells only, as well as positive and negative controls. In addition to exporting the metrics as numerical data for further analysis, it is possible to export individual images or generate image stacks in the form of time-lapse videos with or without the analysis filters (masks) displayed.

11. "Raw" metrics from the assay analysis can be further analyzed using an appropriate graphical analysis tool, for example, Prism (GraphPad, CA).

3.4 xCELLigence Killing Assay Protocol

1. Prior to setting up a T cell-mediated killing assay, perform a QC of the system using resistor plates placed in each cradle of the RTCA Station, in accordance with the manufacturer's instructions, to ensure proper functioning of the equipment during the assay [13, 16].

2. Perform a background reading of the assay plate: Add 50 μL of assay media to an E-Plate and equilibrate to 37 °C for at least 1 h before the measurement. Fill gaps between the wells with PBS to reduce evaporation. The measurement is performed by setting the software to 1 sweep (measurement) at a 1-min interval.

3. Seed target cells at the seeding density predetermined in preliminary experiments (using the protocol in Subheading 3.2) in 150 μL, equilibrate at room temperature for 30 min, and place assay plate in the RTCA Station.

4. Prepare PBMC using the protocol in **Note 2**.

5. Enumerate recovered effector cells, seed in an appropriate tissue culture flask at 1×10^6 cells/mL, and incubate

overnight. This step is essential when using PBMCs to deplete myeloid cells (which will adhere to the culture flask), as the presence of any additional cell types with the potential to adhere to the 96-well E-Plate surface will affect measured impedance and complicate interpretation of the data. Note that this step is omitted if using purified T cell subsets, which should be prepared independently on the day of the assay and added to the assay plate at **step 10**. The number of effector cells put in culture should be enough to allow for 50% death overnight. If labeling the effector cells to enable multiplexed analysis of T cell proliferation (*see* **Note 7**), we have found increasing the number of effectors by a further 50% is important to account for loss in the staining process.

6. After 24 h, remove the E-Plate containing target cells from the RTCA Station and carefully and accurately remove 100 μL of the medium from each well.

7. Prepare a stock solution of soluble test agent(s), for example, T cell redirecting ImmTAC or sTE.

8. Recover effector cells growing in suspension and pass through a pre-wet 30 μm separation filter.

9. Enumerate recovered effector cells.

10. Prepare a concentrated stock of effectors.

11. Remove E-plate from RTCA Station and pipette 50 μL of test agent, followed by 50 μL of effector cells into appropriate wells of assay plate. Use assay media to equilibrate all wells to a total volume of 200 μL.

12. As a nonspecific positive control, addition of 0.4% Triton-X will result in efficient (100%) cell lysis. Negative control wells consisting of target cells alone are essential for subsequent normalization.

13. Return the assay plate to the RTCA Station and set to record CI measurements, taken every 2 h over an additional 72 h.

14. At the end of the assay, the data can be analyzed using associated xIMT software.

15. Normalize impedance, expressed as a CI value to a relevant time point, for example, the point of sTE molecule addition.

16. The normalized values can then be used to calculate % cytolysis, a metric expressing the level of measured impedance relative to untreated target cells. To ensure successful calculations, each plate must contain negative controls containing only targets and targets with effectors (no drug or sTE), as well as a positive control (known as 100% lysis) with the highest concentration of the test drug or a nonspecific inducer of target cell death.

17. xIMT software contains additional built-in analysis functions, including calculations of EC_{50} (half maximal effective concentration), IC_{50} (half inhibitory concentration), or KT_{50} (time to reach 50% of killing).

4 Results

4.1 IncuCyte Redirected T Cell Killing Assay

To evaluate ImmTAC-mediated concentration-dependent redirected T cell killing using the IncuCyte platform, we seeded a pancreatic cancer cell line expressing the ImmTAC's target peptide-HLA complex (pHLA) at a predetermined optimal cell density (Fig. 1a, Subheading 3.2). After 24 h, the target cancer cells were treated with an ImmTAC molecule at a range of concentrations and the T cell killing assay initiated by coculturing with HLA-matched PBMC in assay media containing a caspase-3/7-sensitive fluorescent probe. A process definition was then applied to exclude green fluorescent signal originating from apoptotic T cells. Analysis of the green object count, a readout of induced target cell apoptosis, revealed induction of target cell death in cocultures treated with a moderate affinity ImmTAC, but not in cocultures without ImmTAC or in monocultures of target cancer cells or PBMC (Fig. 2a). Importantly, the magnitude and kinetics of redirected T cell killing were ImmTAC concentration-dependent. Comparison of the rate of killing observed with an ImmTAC molecule with moderate affinity for target pHLA to that of different ImmTAC molecule that had been engineered further to exhibit high affinity for target pHLA (Fig. 2b) revealed different kinetics and dose effects. These differences highlight some features of redirected T cell killing data that can cause complications in interpreting kinetic data of this kind. At the highest concentrations of ImmTAC tested, the magnitude of the apoptotic target signal appeared reduced in comparison to lower concentrations. This effect is a consequence of two parameters. Firstly, IncuCyte software quantifies apoptosis on a temporal basis and is not cumulative. Secondly, when the rate of T cell killing exceeds the rate of target cell proliferation, the pool of target cells available becomes diminished in the early stages of the assay resulting in a plateau or reduction in amplitude of killing in the later stages, as the pool of available target cells is smaller. Consequently, higher rates of killing can appear worse than lower rates of killing. This effect can be visualized from the images acquired during the assay, but it can also be mitigated by using target cells that are fluorescently labeled (*see* **Note 4**).

To demonstrate the utility of labeled target cells, we performed a killing assay on target cells stably expressing a commercially available nuclear restricted mKATE2 fluorescent protein, Nuclight Red (Sartorius). For killing assays using two-color fluorescence, the IncuCyte was set up in accordance with the manufacturer's

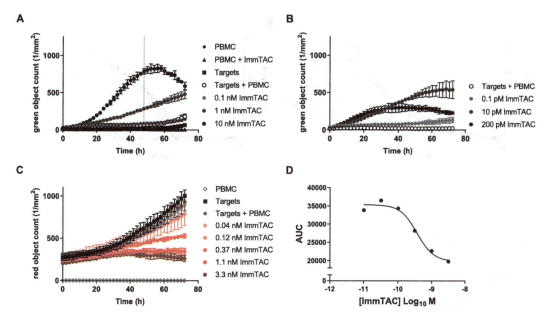

Fig. 2 ImmTAC-mediated T cell redirected cell killing quantified using the IncuCyte platform. Cocultures of PBMC and a target cancer cell line were treated with either a moderate affinity (**a**) or high affinity (**b**) ImmTAC in the presence of a fluorescent caspase 3/7-sensitive probe for 72 h and imaged using an IncuCyte ZOOM kinetic live cell imaging system equipped with a 10× objective at 2-h intervals. Target cell killing (apoptosis) was detected by measurement of green fluorescence, analyzed using the IncuCyte ZOOM software and the green object metric to exclude signal originating from apoptotic PBMC effector cells. (**c**) Cancer cells stably expressing Nuclight Red (nuclear-restricted mKate2 fluorescent protein) were treated with ImmTAC and cocultured with PBMC in an IncuCyte killing assay. Target cell proliferation was detected by measurement of red fluorescence cell nuclei, analyzed using the IncuCyte ZOOM software and the red object metric. (**d**) Area under the curve analysis (AUC) of data generated from the killing assay described in (**c**). Error bars indicate S.D. of replicate samples

instructions and technical notes, which provide practical recommendations on additional controls and spectral unmixing [14]. To quantify killing of the target cells in the assay, a process definition was applied to count red fluorescent signal originating from the nuclei of viable target cells stably expressing Nuclight Red. Analysis of the red target cell object count revealed a reduction in the number of target cells and a reduction in the rate of target cell proliferation in cocultures treated with ImmTAC, but not controls (Fig. 2c). Target cell count data can be readily converted into a dose-effect curve using area-under-the-curve analysis in suitable software such as Prism (GraphPad) (Fig. 2d).

4.2 xCELLigence Redirected T Cell Killing Assay

To compare and contrast quantification of redirected T cell killing using the xCELLigence platform, we again seeded a pancreatic cancer cell line expressing the ImmTAC target pHLA at a predetermined optimal cell density (Fig. 1b, Subheading 3.2). After 24 h, the target cancer cells were treated with ImmTAC molecules at a range of concentrations in a label-free coculture with PBMC HLA-matched to target pHLA.

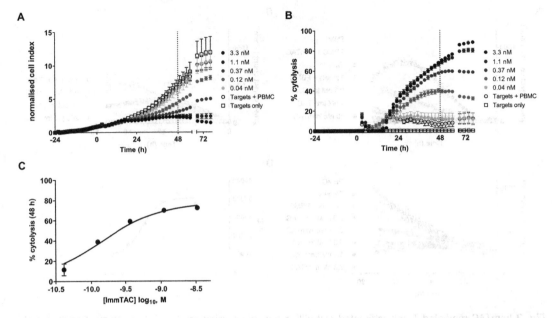

Fig. 3 ImmTAC-mediated T cell redirected cell killing quantified using the xCELLigence platform. Target cancer cells were cultured for 24 h, prior to addition of monocyte-depleted PBMC and treatment with T cell redirecting ImmTAC (0 h). Cocultures were incubated in a label-free assay system, and electrical impedance in each assay well was measured at 2-h intervals using the RTCA xCELLigence platform. (**a**) Target cell killing was detected by relative loss of electrical impedance in comparison to target cells cultured alone, expressed as Cell Index (CI). (**b**) CI values were then normalized to the CI value measured at the last time point before ImmTAC addition and converted into % cytolysis using xIMT software. (**c**) % cytolysis at 48 h plotted as a function of ImmTAC concentration. Error bars indicate S.E.M. of replicate samples

To quantify ImmTAC-mediated redirected T cell killing, we first normalized the measured CI values to the time point at which the ImmTAC molecule was added (24 h) (Fig. 3a). Using a built-in function of the xCELLigence xIMT software, the normalized values were used to calculate % cytolysis, a metric useful for comparing and plotting effects dependent on ImmTAC concentration (Fig. 3b). We sought to interpret relative changes in CI against concentration in order to assess the efficacy of the ImmTAC molecule. For this, % cytolysis data at 72 h (48 h after the time point of ImmTAC molecule addition) was further analyzed as a function of ImmTAC concentration using nonlinear fit regression (GraphPad Prism v 7.02 (GraphPad Software, La Jolla, California USA, www.graphpad.com) to obtain an EC_{50} value for the ImmTAC molecule tested (Fig. 3c).

5 Notes

1. Comparison of IncuCyte and xCELLigence Methodologies
   A summary of the main benefits of each platform can be found in Table 1. The relative merits of each platform depend on the nature of the research, the question being addressed, and the intended use of the data.

2. Effector Cells
   PBMC can be obtained from numerous commercial suppliers, available as convenient-to-use, cryofrozen aliquots. In comparison to purified CD8+ T cells, PBMC better recapitulate the complexities of the immune system in vitro because of the presence of multiple cell lineages.

Table 1
Comparison of main features of each assay platform

Function	IncuCyte	xCelligence
Data acquisition	Real time	Real time
Capacity	Six micro-titer plates (384 and 96-well)	Six micro-titer plates (96-well) or four 384 plates (if equipped with robotic arm and multiple HT RCTA units)
Scan speed	Minutes (dependent on the number of samples and channels)	15 s
Capability	Adherent and nonadherent targets	Adherent and nonadherent (limited to specific cell lineages) targets
Ability to visualize killing of target cells directly	Yes	No
Assay format	Best results require use of fluorescent probes or dyes	Label free
Amenable to multiplexing with additional endpoint assays	Yes	Yes
Possibility to coculture multiple target cells types in same well	Yes (S3 equipped with cell-by-cell analysis module)	No
Data and analysis	Qualitative and quantitative. Image-based analysis can involve a degree of subjective discrimination between target and effector cells	Highly quantitative and straight forward—based on impedance. Software includes automatic calculation of time-dependent EC_{50}

Remove cells from liquid N_2 and transport on dry ice prior to thawing at 37 °C in a water bath.

(a) Immediately after the cells have thawed, transfer effectors to an empty tube and wash by addition of pre-warmed media dropwise. Centrifuge effectors at 300 × g for 5 min, remove supernatant, and resuspend in pre-warmed media.

(b) Pass the effector cell suspension through a pre-wet 30 μm separation filter and enumerate recovered effector cells. We have found accurate quantification, for example, by using AOPI staining to determine a viable cell count and understand live/dead ratios, to be beneficial.

(c) Adjust volume to the relevant density required in the assay. An E:T ratio of 10:1 is optimal for use in T cell redirected killing assays using soluble ImmTAC reagents and PBMC. Empirical determination of E:T ratio may be necessary for other therapeutic modalities.

In our experience, the protocols described in this chapter are easily modified to substitute PBMC for subpopulations of T cells. For example, using standard techniques, we have validated the approach described here using CD8+ T cells isolated from peripheral blood lymphocytes obtained from healthy human volunteers using Ficoll-Hypaque density gradient separation (Lymphoprep, NycomedPharma) and magnetic bead immunodepletion (MiltenyiBiotec) to remove CD19+, CD4+, and CD14+ cells [10, 17]. As a guide, a 1:1 or 2:1 E:T ratio using purified CD8+ T cells is comparable to a 10:1 E:T ratio using PBMC. Different T cell subsets vary in their ability to kill target cells [18], therefore, optimizing E:T ratio is recommended.

Alternative Target Cells

3. Nonadherent Cells

It is possible to adapt the protocols described here for the assessment of redirected T cell killing of nonadherent target cells. Nonadherent target cells can be tethered to the surface of xCELLigence or IncuCyte assay plates by pre-coating plates either with antibodies specific for membrane proteins expressed on the surface of the target cell, with extracellular matrix components, or with the synthetic amino acid poly-L-ornithine (PLO) [14, 15]. Incubating the plate for an additional 30 min at room temperature after cell seeding and before placing in the incubator is essential to promote an even distribution of nonadherent target cells settling to the bottom of the assay plate wells.

4. Fluorescently Labeled Cells

   The IncuCyte platform is equipped to detect both green and red fluorescence (Green fluorescence channel excitation: 440–480 nm, emission: 504–544 nm; Red fluorescence channel excitation: 565–605 nm, emission: 625–705 nm). Using fluorescently labeled cells has the potential to aid interpretation of data as well as augment the data set. There are a number of considerations to this approach:

   (a) A key consideration is whether to label the target or the effector cell population. Labeling the effector population for imaging may preclude or impair assessment of proliferation using CFSE-based dyes (*see* **Note 7**). Additionally, because activation of T cells triggers proliferation, dye is readily lost from the effector population. For this reason, we have found that the best results are obtained by labeling the target cells, as they tend to proliferate less rapidly.

   (b) The choice of label is also an important consideration. Sartorius has produced useful technical notes and a white paper to help users evaluate the choice between using a fluorescent dye or expression of a fluorescent protein to label target cells [14, 19].

   (c) Generating cell lines that stably express a fluorescent protein avoids the problem associated with loss of fluorescent signal when cells divide. This has the added value of augmenting the data set to include a quantitative assessment of the number of cells at a given time point as well as improving normalization.

5. Normal or Primary Cells

   In addition to assessing the potency of therapeutic agents capable of eliciting or redirecting T cell killing of immortalized cancer cell lines, it is theoretically possible to adapt the protocols described here either for assessment of potency using primary cancer cells or to make safety assessments by replacing cancer cells in the assay with primary cells isolated from healthy donors to represent normal cells.

   In both instances, a key consideration is the survivability of the target cells in the assay media. Because the assays are essentially a readout of T cell response, we recommend using assay media that reliably and robustly maintain T cell viability and function. Primary tumor and normal cells may have limited capacity to survive and proliferate in vitro without complex media. However, this type of media is frequently supplemented with factors that have the potential to alter T cell function (for example, corticosteroids) or are detrimental to T cell viability. Systematic assessment of the ability of the target primary cell to survive (and proliferate) in assay media using the protocol

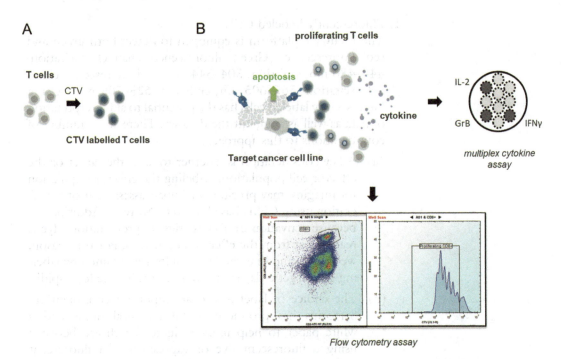

Fig. 4 Schematic diagram presenting overview of multiplexing (**a**) Cartoon depicting pre-labeling of T cell effectors with cell trace violet (CTV) dye to measure T cell proliferation in a multiplexed killing assay. (**b**) The methods described in this chapter are readily amenable to multiplexing with additional readouts of T cell activation, for example, by measurement of cytokines released into killing assay coculture supernatants and analyzing T cell effector proliferation

described in Subheading 3.2 is therefore recommended. In addition, it may be necessary to assess T cell function, viability, and proliferation if using alternative assay media capable of supporting primary cell growth.

6. Multiplexed Endpoint Assays
The live-cell/non-lytic nature of the redirected T cell killing assays described in this chapter makes them amenable to combination or multiplexing with additional assays. Multiplexing affords the opportunity to augment the data derived from a single biological experiment (Fig. 4). In particular, we have found collecting cell culture media and/or responding effector cells to evaluate the extent of T cell activation by measurement of either soluble factors or cell proliferation to be particularly impactful.

7. Multiplexed T Cell Proliferation Protocol
A key consideration in experimental design and setup is to account for depletion of the effectors during the CFSE/CTV staining process. We estimate that this can result in a loss up to 30% of viable cells during the CTV staining.

(a) Prior to addition of PBMC or purified CD8+ or CD4+ T cells to the assay plate, wash cells once with sterile PBS, pellet in a centrifuge at 300 × g, and resuspend to a concentration of 5×10^6 cells/mL.

(b) Prepare a 5 mM CTV stock in accordance with manufacturer's recommendation and protect from light.

(c) Add CTV stock to effector cell suspension at final concentration of 5 μm, resuspend thoroughly but gently using manual pipetting, and incubate at 37 °C for 20 min. It is important to retain an aliquot of effectors that are unlabeled for use as compensation controls.

(d) Add 5 volumes of assay media to quench the dye and harvest cells at 300 × g for 5 min.

(e) Resuspend the cells at the desired concentration for addition to the assay plate. We have found it pertinent to recount the cells rather than rely on the cell count prior to CTV staining to account for cells lost during staining.

(f) Set up the assay (follow protocol from Subheadings 3.3 and 3.4 for InCucyte and xCELLigence assays, respectively).

(g) After 96 h of assay initiation, remove assay plate from the reader and carefully but thoroughly resuspend effector cells (which will have settled at the bottom of the well) into the culture supernatant using gentle manual pipetting and transfer to a U-bottom 96-well plate.

(h) Pellet the cells using centrifugation at 350 × g for 3 min and discard the supernatant.

(i) Wash cell pellet with PBS to remove serum and resuspend in 50 μL of antibody mix in PBS supplemented with antibodies and live/dead stain:
   - Anti CD8—APC 1:50.
   - Anti CD4—PE-Cy7 1:100.
   - Anti CD3—APC-Cy7 1:100.
   - Zombie Green 1:500.

(j) Compensation controls should be resuspended in PBS supplemented with individual antibodies using cells that have not been labeled with CTV.

(k) Incubate for 30 min at 4 °C and protect from light.

(l) Harvest cells by centrifugation and wash with 50 μL FACS buffer. Repeat wash step.

(m) Cells can be fixed at this stage after staining using a solution of 2% PFA/PHEM or a commercially available fixative at 37 °C for 15 min. At the end of the incubation, wash the cells with FACS buffer, spin the plate at 300 × g for 3 min, and resuspend in PBS or FACS buffer for storage at 4 °C. On the day of data acquisition, spin plate down and resuspend in fresh FACS buffer.

We have successfully used either Mesoscale Discovery (MSD) electrochemiluminescence ELISA or multiplexed bead assays, for example, QBead assay using Intellicyt®iQue Screener PLUS (Sartorius) or Luminex, to identify and analyze the levels of soluble cytokines secreted by effector cells in the context of redirected T cell killing assays.

Depending on the nature of assay used, therapeutic test agent, cytokines of interest, and method of detection, preliminary experiments may be necessary to determine the optimal time point to sample the cell culture supernatant. Although we have found that most cytokines can be reliably detected at 24–48 h, it is prudent to determine an optimal sampling time point in preliminary time course studies.

(a) The setup of the assay needs to be adjusted to increase the total volume in wells by addition of an extra 50–100 μL media to account for supernatant that will be removed for cytokine analysis.

(b) At the appropriate time point, carefully remove 10–60 μL supernatant from each well of the IncuCyte or xCELLigence killing assay into a V-bottom polypropylene 96-well plate.

(c) Briefly centrifuge to remove cells or debris from the supernatant and transfer to a fresh V-bottomed 96-well plate.

(d) Seal plate and store at −80 °C until ready for analysis.

6 Concluding Remarks

Interest in understanding the detailed mechanisms involved in the immune response to cancer, and in particular immune tolerance, is highlighted by recent therapeutic advances in the field of I/O. This has driven a need for better cell-based in vitro models to enable academic and industrial researchers to understand and manipulate immune tolerance of cancer and/or promote immune-mediated killing of tumor cells.

Here, we describe protocols using two platform technologies capable of evaluating kinetic redirected T cell killing over longer time periods than can be achieved using traditional methods. Using the methods described in this chapter, we show that both the IncuCyte and xCELLigence platforms can generate robust and reproducible data to assess redirected T cell killing. Both platforms are also readily amenable to multiplexing with additional readouts of T cell response, including cytokine release and T cell proliferation.

Each platform offers distinct advantages (Table 1) and we suggest that they are complementary, rather than competing technologies. The relative merits of each platform will depend on the ultimate purpose and use of the data. In our experience, the IncuCyte is invaluable for mechanism of action studies and in early discovery because of the benefits offered by image-based analysis. In contrast, we tend to exploit the xCELLigence platform in later stage drug discovery programs to either make comparative and temporal $EC_{50/x}$ measurements or to quantify ImmTAC redirected T cell killing activity in combination with other therapeutic agents. Collectively, these data can inform on the therapeutic potential and the selection of a clinical safe starting dose based on a minimal anticipated biological effective level (MABEL) for an ImmTAC molecule during preclinical development [1].

As such, both basic research and preclinical drug development of fully human-specific therapies such as sTE or TCR-based cell therapies can benefit from the robust in vitro kinetic killing assay methods described here.

Acknowledgments

We would like to thank Namir Hassan, Jane Harper, Giovanna Bossi, Martina Canestraro, Zoe Donnellan, and all who have helped establish and develop redirected T cell killing assays at Immunocore Ltd. We would also like to thank Michelle McCully for critical review and help with the preparation of this manuscript and thank Sartorius and ACEA BioSciences for their technical support. Tomasz Dobrzycki and Andreea Ciuntu contributed equally to this work.

References

1. Harper J, Adams KJ, Bossi G, Wright DE, Stacey A, Bedke N, Buisson S, Martinez-Hague R, Blat D, Humbert L, Buchanan H, Le Provost GS, Donnellan Z, Carreira R, Paston SJ, Weigand LU, Canestraro M, Sanderson JP, Botta Gordon-Smith S, Lowe KL, Rygiel KA, Powlesland AS, Vuidepot A, Hassan NJ, Cameron BJ, Jakobsen BK, Dukes J (2018) An approved in vitro approach to preclinical safety and efficacy evaluation of engineered T cell receptor anti-CD3 bispecific (ImmTAC) molecules. PLoS One 13(10):e0205491

2. Chen DS, Mellman I (2013) Oncology meets immunology: the cancer-immunity cycle. Immunity 39:1–10

3. Wierda WG, Mehr DS, Kim YB (1989) Comparison of fluorochrome-labeled and ^{51}Cr-labeled targets for natural killer cytotoxicity assay. J Immunol Methods 122:15–24

4. Baumgaertner P, Speiser DE, Romero P, Rufer N, Hebeisen M (2016) Chromium-51 (^{51}Cr) release assay to assess human T cells for functional avidity and tumor cell recognition. Bio-protocol 6(16):e1906. https://doi.org/10.21769/BioProtoc.1906.

5. Sartorius (2018) Live-cell analysis handbook: a guide to real-time live-cell imaging and analysis, 2nd edn. Bohemia, NY, Sartorius

6. ACEA Biosciences, Inc. (2013) ThexCELLigence System. ACEA Biosciences, Inc., San Diego, CA

7. Hassan NJ, Oates J (2013) The T cell promise. Eur Biopharm Rev, Magazine 12, issue 200 (Summer 2013 edition).

8. Lowe K, Cole D, Kenefeck R, OKelly I, Lepore M, Jakobsen BK (2019) Novel TCR-based biologics: mobilising T cells to warm 'cold' tumours. Cancer Treat Rev 77:35–43

9. Kontermann RE, Brinkmann U (2015) Bispecific antibodies. Drug Discov Today 20:838–847

10. Oates J, Hassan N, Jakobsen B (2015) ImmTACs for targeted cancer therapy: why, what, how, and which. Mol Immunol 67:67–74

11. Liddy N, Bossi G, Adams KJ, Lissina A, Mahon TM, Hassan NJ, Gavarret J, Bianchi FC, Pumphrey NJ, Ladell K, Gostick E, Sewell AK, Lissin NM, Harwood NE, Molloy PE, Li Y, Cameron BJ, Sami M, Baston EE, Todorov PT, Paston SJ, Dennis RE, Harper JV, Dunn SM, Ashfield R, Johnson A, McGrath Y, Plesa G, June CH, Kalos M, Price DA, Vuidepot A, Williams DD, Sutton DH, Jakobsen BK (2011) Monoclonal TCR-redirected tumor cell killing. Nat Med 18:980–987

12. Sato T, Nathan PD, Hernandez-Aya L, Sacco JJ, Orloff MM, Visich J, Little N, Hulstine A-M, Coughlin CM, Carvajal RD (2018) Redirected T cell lysis in patients with metastatic uveal melanoma with gp100-directed TCR IMCgp100: Overall survival findings. J Clin Oncol 36:9521

13. ACEA Biosciences, Inc (2018) xCELLigence® RTCA handbook: cancer immunotherapy. ACEA Biosciences, Inc., San Diego, CA

14. https://www.essenbioscience.com/en/resources/documents/

15. Obst R (2015) The timing of T cell priming and cycling. Front Immunol 6:563. https://doi.org/10.3389/fimmu.2015.00563

16. https://www.aceabio.com/resources/rtca-resources/

17. McCormack E, Adams KJ, Hassan NJ, Kotian A, Lissin NM, Sami M, Mujic M, Osdal T, Gjertsen BJ, Baker D, Powlesland AS, Aleksic M, Vuidepot A, Morteau O, Sutton DH, June CH, Kalos M, Ashfield R, Jakobsen BK (2013) Bi-specific TCR-anti CD3 redirected T cell targeting of NY-ESO-1- and LAGE-1-positive tumors. Cancer Immunol Immunother 62:773–785

18. Boudousquie C, Bossi G, Hurst JM, Rygiel KA, Jakobsen BK, Hassan NJ (2017) Polyfunctional response by ImmTAC (IMCgp100) redirected CD8$^+$ and CD4$^+$ T cells. Immunology 152:425–438

19. Trezise DJ, Campwala H, Appledorn D, Rauch J, Roddy M, Dale TJ (2017) Fluorescent cell-labeling strategies for IncuCyte® live-cell analysis. Why? What? And When? White paper. Essen BioScience, a Sartorius company, Ann Arbor, MI

Chapter 5

Induction and Potential Reversal of a T Cell Exhaustion-Like State: In Vitro Potency Assay for Functional Screening of Immune Checkpoint Drug Candidates

Eden Kleiman, Wushouer Ouerkaxi, Marc Delcommenne, Geoffrey W. Stone, Paolo Serafini, Mayra Cruz Tleugabulova, and Pirouz M. Daftarian

Abstract

Tumor infiltrating lymphocytes (TIL) entering the tumor microenvironment (TME) encounter many suppressive factors resulting in a spectrum of possible differentiation paths, many with overlapping characteristics. This differentiation spectrum includes T cell anergy, unresponsiveness, quiescence, tolerance, dysfunction/suppression, and exhaustion, each with many subtypes. Immune checkpoint blockade (ICB) cancer drug therapies have focused on reinvigorating exhausted T cells (T_{EX}) and dysfunctional T cells (herein referred to as T_{EX}). Many factors have been attributed to as causative to this exhausted state including sustained surface expression of multiple co-inhibitory receptors, altered transcription factor expression, epigenetic rewiring, and dysregulated metabolism. Antigen persistence is necessary for driving T_{EX} maintenance in both the chronic viral infection setting and cancer. In addition to persistent antigen exposure, TILs within the TME encounter numerous tumor-mediated immunosuppressive metabolic by-products, suppressive cytokines, hypoxia, and cellular debris which converge to suppress T cell function and uniquely alter its transcription factor profile. These suppressed or dysfunctional T cells are incapable of mounting an optimal anti-tumor response in part due to lack of fitness in competing for glucose and oxygen. This chapter describes two different potency assays: (a) first we describe optimization of a core recall antigen-based in vitro potency assay for screening of immunopotentiating drug candidates; (b) the second assay bundles the core assay with further addition of immunosuppressive agents derived from the TME making a customizable T cell exhaustion-like assay. Our recall antigen assay is being used as a tool for function-based screening of ICB drug candidates. In this assay healthy human Peripheral Blood Mononuclear Cells (PBMCs) are stimulated with peptide(s) and grown in culture for 1 week. Day 4 supernatants are functionally assayed by ELISA for IFN-γ secretion, and cells are assayed on day 7 by flow cytometry for $CD8^+$ or $CD4^+$ T cell expansion by using a single or cocktail of pMHC tetramers. We have shown that in roughly 30% of donor PBMCs, ICB drugs such as pembrolizumab are able to boost both IFN-γ secretion and antigen-specific recall. One potential explanation for this effect is that the observed increase in T cell co-inhibitory receptor expression and presumed co-inhibitory receptor downstream signaling is ameliorated with ICB drugs releasing these T cells from the repressive effects of these co-inhibitory receptors. We have further enhanced our recall assay to more closely resemble the TME by including metabolites and other suppressive agents at concentrations not normally encountered by T cells in a healthy environment. We show one example of this whereby addition of adenosine to the culture results in suppression of antigen-specific T cell expansion without affecting cell viability. Further, we screened several drug candidates on their ability to counteract the negative effects of adenosine and found that one of the drugs was indeed able to restore antigen-specific T cell expansion. Hence, this TME-based recall antigen assay allows for dissection of the effects of TME-based factors on donor-specific T cell response as well as drug candidate screening.

Key words T cell exhaustion, Antigen specific CD8+ T cells, Recall antigen, Co-inhibitory molecules, Immune checkpoint blockades, Immunotherapy, Adenosine, PD-1, LAG-3, TIM-3, Pembrolizumab

1 Synopsis

The goal of cancer immune checkpoint therapies is to cure tumor-specific T cells from dysfunction which is caused by elements from tumor deposits. In this chapter, we briefly describe the process of T cell exhaustion and how it can be harnessed for immune-oncology drug discovery and more specifically to screen immune checkpoint drug candidates. First, the current function-based characterization methods for immunomodulatory drug candidates will be summarized. We have selected three methods which mimic prerequisite parameters of early T cell exhaustion (T_{EX}). Next, we briefly describe an improved recall antigen-based potency assay as detailed elsewhere. Lastly, we describe an in vitro reversible early T_{EX} model, which may be tailored with cancer-induced T cell suppression agents. Here it should be noted that cancer-induced T cell dysfunction is also referred as to T_{EX}, covering a broad spectrum of different molecular T cell pathologies. The term "T cell exhaustion" in this review mainly refers to an early stage of this process. We have devised an in vitro model that recapitulates components of tumor deposits responsible for inducing T_{EX}.

T_{EX} is a phenomenon that describes a T cell differentiation continuum in which there is a hierarchical progressive diminution of effector function caused in part by elevated and sustained expression of multiple co-inhibitory receptors, altered transcription factor expression, epigenetic rewiring, and dysregulated metabolism [1–3]. Furthermore, T_{EX} cells cannot achieve quiescence [2] and can potentially become dependent on the presence of antigen for their maintenance as was shown in the lymphocytic choriomeninigits virus (LCMV) chronic viral infection model [4], whereas antigen persistence was not needed for T_{EX} maintenance in a cancer antigen model system [5]. Severely exhausted cells become refractory to reactivation [6]. One mechanism that mediates this state, in both CD4+ and CD8+ T cells, is persistent exposure to antigen during chronic infection or cancer [1–3, 6, 7]. However, differences in co-inhibitor receptor expression and transcription factor usage distinguish T_{EX} cells that are derived from chronic viral infection versus cancer [8]. In addition to persistent antigen exposure within the tumor microenvironment (TME), tumor-mediated immunosuppressive metabolic by-products, suppressive cytokines, and cellular debris converge to inhibit T cell function [1–3, 6, 8]. This combined with increased competition for glucose and oxygen serves to further dampen T_{EX} anti-tumor activity [9]. One such immunosuppressive agent is potassium. Excessive extracellular potassium released into the TME from dying tumor cells can reach a concentration that leads to intracellular potassium T cell influx causing them to become dysregulated [10]. This phenomenon leads to decreased T cell uptake and consumption of nutrients, ultimately resulting in reduced effector gene expression such as IFN-γ [11]. Another immunosuppressive agent that is the focus of intense investigation is adenosine. Hypoxic conditions lead to increased extracellular adenosine which suppresses anti-tumor T cell activity via signaling through cognate receptor A2AR [12]. TME suppressive mediators such as these are prevalent in the solid tumor microenvironment and contribute to the resistance of many tumors to immune checkpoint blockade [13].

2 Current In Vitro Function-Based Potency Assays for Immune Checkpoint Drug Candidates

The field of cancer immunotherapy has had many breakthroughs in the last decade. Not too long ago; the community was pleased with a 5% success rate and the current 20–30% success rate in many indications was a dream. The bar is now much higher, and today's challenges include reversal/preventing T_{EX} as well as identifying and converting non-responders. To this end, it is important to expand our limited knowledge of T_{EX}, understanding the differences between tumor-specific T cells before and after they interact with tumors, and understanding the main drivers and mechanisms of tumor-induced T_{EX}. Persistent antigen stimulation causes sustained overexpression of co-inhibitory molecules which is at least partly responsible for T_{EX} [1, 2]. However, T_{EX} encompasses a much broader spectrum, which can be categorized in many different types (early, late, reversible, non-reversible etc.). Many factors influence the degree of T_{EX} including antigen-rich environments, un-helped T cells, altered metabolism, ions, lipids, the milieu of cytokines, combinations of these as well as additional suppressive factors [1, 2, 6].

The three commonly used potency assays are recall antigen assay (RAA), mixed lymphocyte reaction (MLR) [14, 15], and the staphylococcal enterotoxin B assay (SEB). The core principles of these functional assays for potential immunotherapeutic biologics are as follows: (i) these assays can be optimized as a T_{EX} assay by inducing co-inhibitory molecule expression, (ii) screening assay for active antagonist antibodies to immune checkpoint molecules or agonist antibodies to co-stimulatory molecules with read-outs including antigen-specific T cell expansion, IFN-γ secretion, surface expression of activation markers, or IL-2 secretion in MLR [16], SEB, and RAA, and (iii) modularity in enabling manipulation of culture conditions to include TME metabolites that induce T cell suppression/dysfunction with similar read-outs as in (ii). These assays have been utilized by academia and industry for functional characterization of immunotherapeutic candidates from discovery phase all the way to later stages of the drug development [16–18]. As an alternative to animal studies, in vitro potency assays are ideal non-invasive means for functional screening of and for assessment of critical quality attributes of the drug candidate or lead drug candidates. Such assays are required to be validated for sensitivity and specificity, accuracy, representativeness of the mechanism of action, physiological relevance, reproducibility, and finally predictivity of clinical efficacy [19]. However, since most immune checkpoint drug candidates have unique features and there is not enough data for predictability of clinical efficacy, companies employ multiple assays to ensure that in vitro potency assays may be translated to clinic. To this end, similar assays have been used in submissions to regulatory agencies as evidence of non-clinical pharmacological in vitro potency assays (e.g., FDA BLA # 125554). Here, we will discuss the common features of these three assays, some specific features of each assay, and various readouts of these assays, including an improved functional MHC tetramer-guided assay to increase the signal-to-noise ratio. Some related novel phenotypical or activation marker approaches can also be used as a readout of the potency assays and will also be discussed. Finally, we will discuss why multiple in vitro potency assays should be employed for functional screening of immune checkpoint inhibitor drug candidates and argue that a direct measurement of antigen-specific T cell expansion may be most physiologically relevant.

2.1 Challenges of Current Potency Assays for Immune Checkpoint Drug Candidates

There are some common features of these immune potency assays which make them good candidates for functional characterization of candidate biologics with immunomodulatory effects. They all start with an in vitro stimulation phase, an immune activation step of primary immune cells resulting in increased activation markers, cell proliferation, cytokine secretion, antigen-specific $CD4^+$ T cells, and/or antigen-specific $CD8^+$ T cells, which have been used as readouts alone or in combination. Other common features are immune-related biomarkers that are affected and may reveal the activity of candidates of immune checkpoint blockade. For example, biomarkers ranging from surface activation markers (e.g., CD137, CD69, MHC Class II) to intracellular activation markers (e.g., CD107, Ki67, IFN-γ), and cell proliferation. Another common feature is that they do not use cell lines, which makes them more physiologically relevant. Also, these assays all result in the expression of selected immune checkpoint molecules. The expression of such strong co-inhibitory molecules is associated and at least partly responsible for T_{EX} [1, 2], which has been be used to understand the mechanism of action of active immune checkpoint blockades in reversing the T_{EX}. Although the recall antigen-based potency assay mimics early exhaustion in that there is a broad increase in co-inhibitory receptor expression on expanded antigen-specific CD8$^+$ T cells, this in and of itself does not indicate exhaustion as other factors are involved including sustained expression of these co-inhibitory receptors. However, high expression of multiple co-inhibitory receptors is a prerequisite for T_{EX} and a general mechanism to prevent overactivation. The assay may therefore be tailored and manipulated to serve as a T_{EX}-like in vitro model where the active candidate is expected to further increase recall in response donors or to reverse the process of un-responsiveness being caused by a suppressive agent.

2.2 Mixed Lymphocyte Reaction

The mixed lymphocyte reaction (MLR) stems from T cell alloreactivity, which drives transplant rejection. A high number of responding T lymphocytes have direct antigen specificity to alloantigens; thus, for this assay there is no need for priming. The basis of MLR as a potency assay is that, depending on the host, approximately 1–10% of one's T cells recognize and respond to non-self complexes of peptide-major histocompatibility complex (MHC) (alloreactivity). It is unclear how a T cell receptor (TCR) accomplishes such high specificity for a peptide antigen presented by allo-MHC, though there are some theories [20]. Nevertheless, this assay creates an opportunity to test the effect of immune checkpoint inhibitors for further enhancement of the MLR response. MLR is a classic culture method used for studying the allogenic immune-response in vitro. There are reports that indicate a larger role for MHC class II [21], but in general the MLR results in the expansion of both CD4$^+$ T cells and CD8$^+$ T cells. As a result, this potency assay in principal should have value in functional screening of immune checkpoint drug candidates (ICDCs). Although both CD8$^+$ and CD4$^+$ T cells are engaged, it may be that the MLR assay elicits a strong regulatory CD4$^+$ T cell response, including CD3$^+$ CD4$^+$ HLA-DR$^+$ T cells that express CD25, CTLA-4, CD62L, PD-1, and TNFRII [14]. Thus, interpretation of results may become challenging when the MLR assay is used for the potency assays of immune checkpoint inhibitor drug candidates. The common readouts employed when MLR assays are used for testing the potency of ICDC include either cytokine measurement or a proliferation assay. The MLR has been performed by co-culturing CD4$^+$ T cells of one host with the dendritic cells derived from monocytes (moDCs) of PBMC of a second host. For example, when moDCs from one donor were mixed with CD4$^+$ T-cells isolated from a second donor, IL-2 levels were increased in cultures after 48 h when treated with TIM-3 antagonist [22]. Currently, by using MLR, a direct measurement of peptide-specific T cells is not possible.

2.3 Superantigen-Based Assays

Staphylococcal enterotoxin B (SEB), also called a superantigen, binds to the major histocompatibility complex class II (MHC II) molecules and to Vβ of T cell receptors (TCRs), resulting in the stimulation of both monocytes/macrophages and T lymphocytes. There are reports that intercellular adhesion molecule 1 (ICAM1) and leukocyte function-associated antigen 1 (LFA-1) are also involved in cell activation by superantigens [23, 24]. The superantigens connect TCR with MHC II molecules and in turn ignite intracellular signaling cascades and excessive release of proinflammatory cytokines, all of which result in polyclonal T cell proliferation. As a result, the SEB assay using PBMC is a potential candidate for testing the effect of immunomodulators. Indeed, for the immunopathological condition called superantigen-induced shock, the FDA has approved CTLA4-Ig (abatacept) in addition to NFκB, calcineurin, and mTORC1 inhibitors [25].

2.4 Recall Antigen Assay

The original studies that described T_{EX} were in the lymphocytic choriomeninigits virus (LCMV) infection model, where T cell effector function gradually wanes via persistent antigen stimulation through the T cell receptor (TCR). Indeed, persistent CD8$^+$ T_{EX} cells associated with high expression of PD-1 was also first described in the chronic LCMV-infection mouse model. The in vivo LCMV T_{EX} observed was then shown to be reversible in that targeting PD-1/PD-L1 recovered T cells from the dysfunctional state (refer to [2, 3]). Arguably, the Recall Antigen assay is a physiologically relevant assay and has been used for functional characterization of immune checkpoint blockades (ICB) and co-stimulatory agonists including approved immune checkpoint blockades [16–18, 26–29]. The use of the recall antigen assay to determine the potency of immune checkpoint blockades is based on the facts that (a) persistence of antigen results in limitation or suppression of antigen-specific T cell responses, (b) blockade of co-inhibitory molecules, and (c) activation by co-stimulatory agonists reverses or reduces T cell suppression. Although the optimization of the recall antigen assay for the immune monitoring has been reported [30, 31], however, there is an urgent need for such optimizations and validation of the recall antigen assay utility for functional screening and characterization of immune checkpoint drug candidates and similarly for ability of performing the assay in a high throughput manner. Currently the reported protocol is not standardized and suffers from poor reproducibility and signal to noise. To this end, there is no clear understanding of why this assay is highly donor dependent. We have improved and optimized several parameters of the assay including the stimulation stage, the plate polymer, the culture media, interleukins/milieu of cytokines, and finally improving readouts to include cytokines and measurements of expanded recall antigen-specific T cell clonotypes in using a cocktail of MHC tetramers. Figure 1 depicts an improved recall

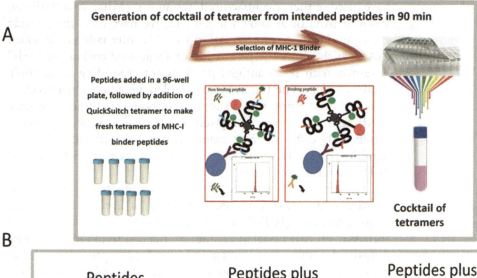

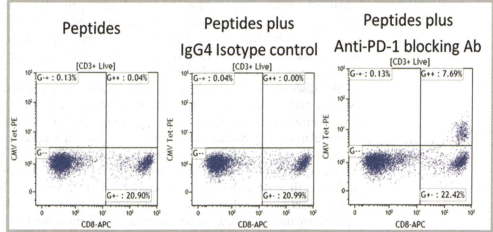

Fig. 1 Improved recall antigen assay for function-based screening of immune checkpoint drug candidates. (**a**) Shows the peptide exchange MHC tetramer platform (MBL International, Woburn MA), which allows the generation of MHC tetramers with intended recall antigen peptide and making a validated cocktail of tetramers corresponding to multiple recall antigens. The tetramers are pooled after individually being made and after desalting step to remove unbound peptides. We have validated this tetramer pool system and the staining pattern was not significantly different from manufactured tetramers for all tested peptides. (**b**) Shows the result of a selected and characterized "PD-1 blockade responder"

antigen assay for function-based screening of immune checkpoint drug candidates. We have selected and characterized donors who have positive recall responses to multiple antigens, for example, those who have low frequency of preexisting CD8+ T cells for CMV, EBV, influenza virus, and WT-1. In doing so, we stimulate with a cocktail of peptides and then detect response with a pool of corresponding tetramers (all may have one fluorochrome). This approach improves the assay's signal-to-noise ratio and better captures the effect of candidate drugs being screened. To this end, we have used either manufactured tetramers (MBL International, Woburn MA) or as shown in Fig. 1a, we have used the peptide exchange MHC tetramer platform (MBL International, Woburn MA), which allows the generation of a validated cocktail of MHC tetramers with recall antigen peptides corresponding to multiple recall antigens. The tetramers are pooled after individual production and a desalting step to remove unbound peptides. We have validated this tetramer pool system and the staining pattern was not significantly different from manufactured tetramers for all tested peptides. Figure 1b shows the result of a selected and characterized "PD-1 blockade responder."

We determined that the assay has the potential to be standardized and used for functional screening of immune checkpoint drug candidates (ICDC) and co-stimulatory agonists. To utilize this assay for functional screening, we needed to (1) make it more industry friendly via a standard protocol, (2) optimize the assay parameters to increase the signal-to-noise ratio, and (3) ensure that the assay is reproducible. To achieve this, we have optimized the assay, so that it recapitulates a chronic infection model to better suit the functional screening of ICDC.

We discovered that healthy donors' PBMC post-stimulation response with a conditioned media may be divided into those whose monocytes express high, medium or low levels of co-stimulatory and co-inhibitory molecules. We then optimized/maximized each element of the assay, the in vitro stimulation as well as the readout of the assay. For example, using an MHC tetramer-guided recall antigen assay in characterized donors, we use CMV, EBV, and influenza-selected peptide pool for stimulation, and as a readout, we used a cocktail of tetramers and activation markers plus interferon-γ secretion. PBMC samples characterized and validated as responders to antigen recall are then used for functional screening of ICDC. We use active immune checkpoint blockades or co-stimulatory biologics that are validated as positive and isotype antibodies as positive and negative controls, respectively. As shown in the Fig. 1, the improved recall antigen assay creates a window of T_{EX}-like and/or suboptimal condition that can be corrected by an active co-inhibitory blockade biologic.

2.4.1 Qualifying the Assay and Validation

Optimizing PBMC-based assays to be performed in a GLP compliance manner has been reported and in general requires additional steps. Since the assay relies on primary cells, we have employed a stepwise approach to assure qualification and validation of the assay. We started by identification of and banking a reasonable size inventory of PBMCs of "responder" donors. We have expanded recall antigen-specific $CD8^+$ T cells and spiked them with syngeneic PBMCs to generate a sizable bank of super responders, and normal responders. We have also characterized the levels of co-inhibitory molecules in tetramer positive populations of each "responder"

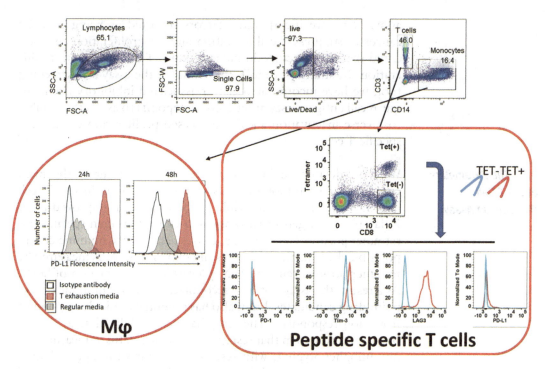

Fig. 2 Induction of co-inhibitory molecules in PBMC of a characterized "PD-1 responder" on day 7 post-recall peptide stimulation

sample, post-recall antigen stimulation (Fig. 2). It should be emphasized that overexpression of co-inhibitory molecules, including PD-1, LAG-3, TIM-3, occurs in both T cell activation and in T cell exhaustion. The data in the resulting spreadsheet allows us to select the potential donors for the drug candidate of interest. We have also validated each batch of our positive controls (PD-1 blockade or LAG-3 inhibitor) and that of our negative controls (IgG4 and IgG1 isotype controls). Data related to these studies are presented in other reports.

3 An In Vitro Reversible Early T Cell Exhaustion Assay

The methods described above are useful, important, and informative for selection of lead drug candidates; however, they can be improved if they include suppressive elements from the TME, mimicking tumor deposits in vitro. Indeed, since the reversal of T cell dysfunction is a hallmark of many cancer immunotherapeutic agents, in vitro selection of drug candidates should also reverse T cell dysfunction. There are many TME elements that render tumor-specific T cells less functional and eventually exhausted. To tailor an in vitro assay system that imitates the cancer-induced suppressive microenvironment, we started with an improved recall antigen assay. Next, we incorporated factors reported as being used by cancer to suppress T cell activation and/or proliferation and validated that these factors do indeed reduce T cell activation/proliferation. In addition, to fully functionalize the assay as a screening tool, we have incorporated cancer-associated inhibitory agents to our recall antigen assay at different timepoints to assess their ability to reverse inactivation and/or decreased proliferation of antigen-specific T cells.

3.1 The Parameters of the T Cell Dysfunction Assay

We have devised an in vitro model that recapitulates the main components of tumor deposits responsible for inducing T cell dysfunction. The core principle of the assay is an improved 6- to 7-day recall antigen in vitro assay to expand antigen-specific and tetramer positive CD8$^+$ T cells. In this assay, active immune checkpoint blockade further enhances the recall response as measured by tetramer positive cells and IFN-γ levels in only a fraction of hosts. As a result, the assay not only recognizes the active immune checkpoint drug candidates, but it also has potential to divide responders from non-responders. The assay has been validated as a PBMC-based culture system that results in a T_{EX}-like status as measured by immunophenotyping, with results consistent with T_{EX} including the expression of co-inhibitory molecules and reduced or limited expansion of peptide-specific T cells and expression of IFN-γ on day 4 post-antigen stimulation. In the process of qualifying the assay, we screened and identified donors whose T cell dysfunction could be reversed by PD-1 blockade and to some degree by inhibitors of LAG-3 and TIM-3. To this end, in our optimized assay system, checkpoint blockade resulted in recovery of T cell responses to peptides shown by two sets of readouts: the percentage of peptide stimulated tetramer positive CD8$^+$ T cells and IFN-γ expression. We next cryopreserved and generated a cell bank of the characterized responders' and non-responders' PBMCs. The PBMCs of characterized responders were used to screen for checkpoint drug candidates or their combinations. The assay is compatible with hybridoma supernatants if the endotoxin levels are less than 4 EU/ml. Figure 3 depicts the steps of reversible T cell dysfunction assay.

An algorithm to simplify the results and capture all readout was also introduced that presents a final "Potency Score" of each drug candidate. This arbitrary scoring system (called T cell dysfunction or exhaustion reversal score) considers IFN-γ and the expansion of tetramer positive T cells by drug candidates (Fig. 4). The algorithm of arbitrary "Potency Score" includes the level of the IFN-γ expression on day 4 post-stimulation (G) and percentages of drug-induced expansion of recall antigen specific tetramer positive CD8$^+$ T cells on day 7 post-peptide stimulation (T), using the following formula, $[G^2 \times T^2]/5$.

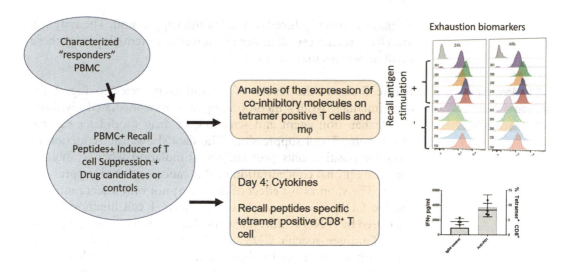

In vitro reversible T cell exhaustion-like Model
Functional screening of immunomodulatory drug candidates that reinvigorate dysfunction antigen specific T cells

Fig. 3 Depicts the process of reversible T cell dysfunction assay

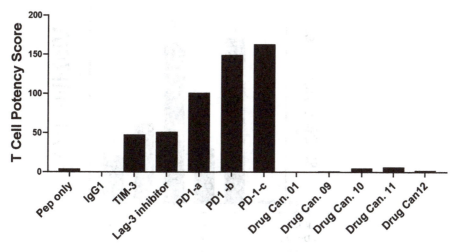

Fig. 4 A function-based T cell potency assay can potentially identify active-drug candidates. T cell "potency score" is calculated by an arbitrary algorithm consisting of the level of the IFN-γ expression on day 4 post-stimulation (G) and the percentage of drug-induced expansion of recall antigen-specific tetramer positive CD8+ T cells (T), in this formula, $[G^2 \times T^2]/5$

3.2 A Plug and Play Assay Model for T Cell Exhaustion Discoveries

We have tested various tumor derived T cell dysfunction agents in this assay system. Here, we will describe one cancer mediated agent that resulted in significant suppression of the expansion of peptide specific tetramer positive CD8+ T cells (or tetramer positive CD4+

T cells) and drug-induced reversal of this suppression. The system is ideal for screening small molecules as well as potential synergy these small molecules may have with ICBs.

3.2.1 Assessment of Tumor-Induced Exhaustion Agents

Figure 5 shows the results of a recall assay (refer to Fig. 1 for experimental layout) incorporating a tumor-mediated dysfunction/exhaustion agent and screening of drug candidates capable of reversing T cell suppression. The data shows the reduction of tetramer positive cells post-antigen stimulation when combined with the optimal concentration of this cancer-induced suppression agent. The suppressive effect of this agent not only affects antigen-specific CD8$^+$ T cell recall but also affects T cell functionality as measured by IFN-γ ELISA (data not shown). Importantly, suppression of antigen-specific CD8$^+$ T cell recall is not due to increased cell death as overall cell viability remained unaltered. This is also true of other dysfunction agents tested (data not shown) for the duration of the recall assay. The suppressive effect of the agent is

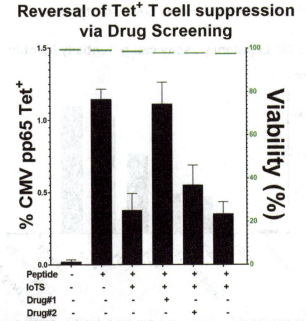

Fig. 5 Data of one undisclosed tumor-mediated inducer of T cell suppression (IoTS) used in a recall antigen assay using healthy donor PBMCs. Black bars represent the percentage of CD8$^+$ T cells that are CMV pp65 tetramer positive (left y axis). The IoTS resulted in a nearly threefold decrease in recall. Two drug candidates are incorporated into the assay to test their ability to reverse T cell suppression. Drug #1 was capable of reversing T cell suppression while drug #2 was not. Vehicle alone control had no effect. Overall cell viability (floating green bars) was not affected by treatment of cells with suppressive agent or drugs (right y axis)

reversible as drug screening assays have identified several candidates capable of reinvigorating T cell antigen recall. Figure 5 illustrates the screening of two drug candidates with one drug demonstrating the ability to reinvigorate antigen-specific T cell expansion.

4 Discussion

The above described in vitro assays result in elicitation of activated T cells (as well as activated antigen presentation cells) and have utility for functional screening of candidate immune checkpoint therapies. Their utility is partially due to a feedback mechanism that includes the expression of co-inhibitory molecules. These assays, however, have some reproducibility and validation challenges. We have further optimized and standardized one of these assays, the recall antigen assay. We selected this assay since it is a controlled antigen system that allows analysis of antigen-specific T cells. Importantly, the system is compatible with tumor-induced T cell dysfunction. The T_{EX} model that was originally described in CMV and LCMV models was later reported to occur using strong T cell stimuli or other antigens in vivo or in vitro [32–34], implying that the immune system has a universal feedback mechanism regardless of the source of the antigen. Furthermore, since we use MHC tetramer to evaluate expansion of T cells, this assay has the potential to be amenable to single cell-based studies. This assay intrinsically is also more amenable to standardization since it allows direct measurement of peptide-specific T cells, which gives a direct cellular readout of the potency of such therapies. Thus, we have focused on further optimization of the recall antigen assay for identifying active candidates of immune checkpoint therapeutics. First, we developed protocols to characterize and cryopreserve a bank of PBMCs from healthy donors for their expression of a panel of immune checkpoint molecules, post T cell and monocyte stimulation. To this end, we have analyzed expression of immune checkpoint molecules for cryopreserved PBMCs, allowing us to select cells that express varying levels of the intended co-inhibitory or co-stimulatory molecules on monocytes or T cells. Using multiple manufactured tetramers or a peptide exchange tetramer platform, with which we have been able to generate tetramers in house amenable to making pooled tetramers to enhance the signal-to-noise ratio resulting in an enhanced readout window for active-drug candidates. A key step in the assay system is to identify active donors to recall antigens whose PBMCs have optimum expression of the checkpoint pathway of interest. Interestingly, PBMCs from characterized donors positive for multiple recall antigens (e.g., CMV, EBV, FLU, and WT-1) may be stimulated with a cocktail of related peptides and the further expansion of T cell clonotypes can be analyzed by a pool of corresponding tetramers. PBMCs of such donors are stimulated with a pool of relevant peptides with either test drug candidates or with positive and negative controls (e.g. PD-1 blockade and matched isotype immunoglobulin). While the main readout is a cocktail of MHC class I tetramers corresponding to peptides that were used in the stimulation phase, the assay has been validated to include additional activation markers (e.g., CD137) and cytokines (e.g., IL-2 and IFN-γ) as additional readouts. The assay system, therefore, generates a greater window of tetramer positive T cells as well as cytokine readouts allowing us to capture small changes caused by candidate therapeutics or their combinations.

Tumor deposits are environments rich with agents that render inhibitory signaling events which have not previously been incorporated into the recall antigen assay. We hypothesized that in vitro functional screening of immunomodulatory drug candidates requires addition of major substances that cancer produces to suppress immune system. In order to improve the assay to better fit the cancer microenvironment, we incorporated tumor-induced suppressive agents to the recall antigen assay. By including these suppressive agents, we have created a recall antigen assay that allows screening of small-molecule drug candidates alone or in combination with current ICIs.

5 Conclusion

The process of immune checkpoint or immunopotentiating drug development is very costly and the selection of lead candidates requires multiple steps. Many characteristics are involved in selection of the lead candidates: chemical property (e.g., stability and potential for scale-up), pharmacological properties (e.g., affinity, function-based potency, and efficacy in animal models), pharmacokinetics (e.g., bioavailability for intended route of administration, stability and half-life, and biodistribution suitable for intended use), and safety and toxicity (e.g., acceptable in animal toxicity studies and immunogenicity) [35, 36]. Generally, as for the pharmacological properties, the first step includes binding assays and affinity assessments to select a range of top binders, which in some cases narrows down the candidates from hundreds to tens of candidates. At this stage, to select the top few most active candidates, pathway specific functional assays with physiological relevance are urgently needed. Here, we have presented a T_{EX}-like/T cell dysfunction assay to identify active immune checkpoint candidates. This assay has the potential to be combined with additional cancer-induced suppression agents to resemble the TME. This TME version of our T cell exhaustion assay has utilities such as testing PD-1 combinations with other drug candidates such as small molecules, gene editing technologies, and novel immunopotentiation candidates.

Acknowledgements

All assays and data were generated at JSR Life Sciences and MBL International. M.C.T. trained as an intern at JSR Life Sciences when she contributed to some studies.

References

1. Wherry EJ, Kurachi M (2015) Molecular and cellular insights into T cell exhaustion. Nat Rev Immunol 15(8):486–499
2. McLane LM, Abdel-Hakeem MS, Wherry EJ (2019) CD8 T cell exhaustion during chronic viral infection and cancer. Annu Rev Immunol 37:457–495
3. Hashimoto M et al (2018) CD8 T cell exhaustion in chronic infection and cancer: opportunities for interventions. Annu Rev Med 69:301–318
4. Shin H et al (2007) Viral antigen and extensive division maintain virus-specific CD8 T cells during chronic infection. J Exp Med 204(4):941–949
5. Schietinger A et al (2016) Tumor-specific T cell dysfunction is a dynamic antigen-driven differentiation program initiated early during tumorigenesis. Immunity 45(2):389–401
6. Yi JS, Cox MA, Zajac AJ (2010) T-cell exhaustion: characteristics, causes and conversion. Immunology 129(4):474–481
7. Crawford A et al (2014) Molecular and transcriptional basis of CD4(+) T cell dysfunction during chronic infection. Immunity 40(2):289–302
8. Thommen DS, Schumacher TN (2018) T cell dysfunction in cancer. Cancer Cell 33(4):547–562
9. Chang CH et al (2015) Metabolic competition in the tumor microenvironment is a driver of cancer progression. Cell 162(6):1229–1241
10. Eil R et al (2016) Ionic immune suppression within the tumour microenvironment limits T cell effector function. Nature 537(7621):539–543
11. Vodnala SK et al (2019) T cell stemness and dysfunction in tumors are triggered by a common mechanism. Science 363(6434):pii: eaau0135
12. Ohta A, Metabolic Immune A (2016) Checkpoint: adenosine in tumor microenvironment. Front Immunol 7:109
13. Anderson KG, Stromnes IM, Greenberg PD (2017) Obstacles posed by the tumor microenvironment to T cell activity: a case for synergistic therapies. Cancer Cell 31(3):311–325
14. Revenfeld ALS et al (2017) Induction of a regulatory phenotype in CD3+ CD4+ HLA-DR+ T cells after allogeneic mixed lymphocyte culture; indications of both contact-dependent and -independent activation. Int J Mol Sci 8:7
15. Ghosh S et al (2019) TSR-033, a novel therapeutic antibody targeting LAG-3, enhances T-cell function and the activity of PD-1 blockade in vitro and in vivo. Mol Cancer Ther 18(3):632–641
16. Wang C et al (2014) In vitro characterization of the anti-PD-1 antibody nivolumab, BMS-936558, and in vivo toxicology in non-human primates. Cancer Immunol Res 2(9):846–856
17. Grenga I et al (2016) A fully human IgG1 anti-PD-L1 MAb in an in vitro assay enhances antigen-specific T-cell responses. Clin Transl Immunol 5(5):e83
18. Li Y et al (2018) Discovery and preclinical characterization of the antagonist anti-PD-L1 monoclonal antibody LY3300054. J Immunother Cancer 6(1):31
19. European Medicines Agency (2016) Guideline on potency testing of cell based immunotherapy medicinal products for the treatment of cancer. European Medicines Agency, London
20. Wang Y et al (2017) How an alloreactive T-cell receptor achieves peptide and MHC specificity. Proc Natl Acad Sci U S A 114(24): E4792–E4801
21. Hundrieser J et al (2019) Role of human and porcine MHC DRB1 alleles in determining the

intensity of individual human anti-pig T-cell responses. Xenotransplantation 26:e12523
22. Haley Laken KM, Murtaza A, de Silva Correia J, McNeeley P, Zhang J, Vancutsem P, Wilcoxen K, Jenkins D (2016) Discovery of TSR-022, a novel, potent anti-TIM-3 therapeutic antibody. EORTC-NCI-AACR Symposium on Molecular Targets and Cancer Therapeutics, Munich
23. Lamphear JG, Stevens KR, Rich RR (1998) Intercellular adhesion molecule-1 and leukocyte function-associated antigen-3 provide costimulation for superantigen-induced T lymphocyte proliferation in the absence of a specific presenting molecule. J Immunol 160 (2):615–623
24. Gjetting T et al (2019) Sym021, a promising anti-PD1 clinical candidate antibody derived from a new chicken antibody discovery platform. MAbs 11(4):666–680
25. Krakauer T (2017) FDA-approved immunosuppressants targeting staphylococcal superantigens: mechanisms and insights. Immunotargets Ther 6:17–29
26. Jones RB et al (2008) Tim-3 expression defines a novel population of dysfunctional T cells with highly elevated frequencies in progressive HIV-1 infection. J Exp Med 205 (12):2763–2779
27. Wang W et al (2009) PD1 blockade reverses the suppression of melanoma antigen-specific CTL by CD4+ CD25(Hi) regulatory T cells. Int Immunol 21(9):1065–1077
28. Lichtenegger FS et al (2018) Targeting LAG-3 and PD-1 to enhance T cell activation by antigen-presenting cells. Front Immunol 9:385
29. Filippis C et al (2017) Nivolumab enhances in vitro effector functions of PD-1(+) T-lymphocytes and leishmania-infected human myeloid cells in a host cell-dependent manner. Front Immunol 8:1880
30. Li Pira G et al (2008) Evaluation of antigen-specific T-cell responses with a miniaturized and automated method. Clin Vaccine Immunol 15(12):1811–1818
31. Duffy D (2018) Standardized Immunomonitoring: separating the signals from the noise. Trends Biotechnol 36(11):1107–1115
32. Bucks CM et al (2009) Chronic antigen stimulation alone is sufficient to drive CD8+ T cell exhaustion. J Immunol 182(11):6697–6708
33. Eikawa S, Mizukami S, Udono H (2014) Monitoring multifunctionality of immune-exhausted CD8 T cells in cancer patients. Methods Mol Biol 1142:11–17
34. Balkhi MY et al (2018) YY1 upregulates checkpoint receptors and downregulates type I cytokines in exhausted, chronically stimulated human T cells. iScience 2:105–122
35. Hefti FF (2008) Requirements for a lead compound to become a clinical candidate. BMC Neurosci 9(Suppl 3):S7
36. Smith SG et al (2017) Assay optimisation and technology transfer for multi-site immunomonitoring in vaccine trials. PLoS One 12 (10):e0184391

Chapter 6

A Human In Vitro T Cell Exhaustion Model for Assessing Immuno-Oncology Therapies

Lynne S. Dunsford, Rosie H. Thoirs, Emma Rathbone, and Agapitos Patakas

Abstract

T cell exhaustion is central in the pathology of cancer and chronic viral infections. This chapter describes an in vitro method to generate human exhausted T cells. Chronic T cell activation is reconstructed by repeated stimulation with CD3-/CD28-targeting antibodies that results in sustained upregulation of several phenotypic hallmarks of exhaustion including co-expression of PD-1, LAG-3, TIM-3, CTLA-4, and the transcription factor TOX. This phenotype was associated with gradual functional impairment that was characterized by diminished cytokine production, proliferative capacity, and cytotoxic potential. PD-1 blockade could partially restore the functionality of the cells, but not to the levels observed in non-exhausted T cells. This model effectively emulates the T cells of the tumor microenvironment and is applicable for the assessment of various targeted therapeutics including bispecific molecules and combination therapies.

Key words T cell exhaustion, Immunotherapy, Combination therapies, Immuno-oncology, In vitro, Checkpoint inhibitors

1 Introduction

T cell exhaustion is an acquired state of dysfunction that has been linked to the pathology of chronic viral infections and cancer. Activation of naïve T cells is followed by extensive alteration of their molecular circuitry, which in conditions of limited antigen availability results in development of long-lived, self-renewing memory T cells that provide effective immunity [1]. However, in situations of persistent antigen exposure as in cancer and chronic viral infections, an alternative differentiation pathway prevails, which leads to a distinct T cell state characterized by functional impairment [2]. Exhausted T cells (T_{EX}) are characterized by a gradual loss of cytokine production; high concomitant expression of checkpoint inhibitors (e.g., PD-1, LAG-3, and TIM-3); distinct transcriptional, metabolic, and epigenetic profile; and reduced proliferative and survival capacity [3–10].

Therapies targeting checkpoint molecules (viz., PD-1, PD-L1, and CTLA-4) have greatly improved clinical outcomes in cancer; however, only a subset of patients achieve long-lasting benefits. This has been partially attributed to the compensatory immunoregulatory functions provided by other checkpoint molecules [11] (e.g., LAG-3, TIM-3, TIGIT) or other immunosuppressive pathways (e.g., metabolic starvation, regulatory T cells), resulting in a race to develop targeted therapeutics against one or more of these pathways [12–14].

Several in vitro assays are employed to support the development of molecules that target T cell exhaustion (e.g., mixed lymphocyte reaction (MLR), superantigen stimulation, or antigen-specific recall assays). These assays usually incorporate a T cell stimulus (polyclonal or antigen specific) that results in activation, cytokine production, and proliferation of T cells that is enhanced in the presence of immune checkpoint blockade (ICB) treatment. These models, albeit very informative, are conventional T cell activation assays, employing fully functional T cells, that rely on the transient upregulation of checkpoint molecules but lack the physiological relevance of an assay that employs exhausted T cells. Traditionally, T cell exhaustion has been investigated either by using animal models (e.g., lymphocytic choriomeningitis virus (LCMV)) or by analyzing tissue-infiltrating lymphocytes (TILs). These models more accurately emulate the T cell exhaustion phenomenon; however, they are hampered by high cost, sample availability, and species incompatibilities. Thus, there is an unmet need for a convenient, physiologically relevant, and robust T cell exhaustion assay for the functional characterization of checkpoint blockade treatment.

This chapter describes a recently developed in vitro model for the generation of human exhausted T cells [4] that can be employed in secondary assays, such as MLR and cytotoxicity assays. The model is based on repeated stimulation of isolated T cells from healthy donors with CD3/CD28 Dynabeads® (Fig. 1) that results in phenotypic and functional exhaustion. The in vitro-generated exhausted T cells co-express multiple co-inhibitory receptors, including TIM-3, LAG-3, PD-1, and CTLA-4, and upregulate thymocyte selection-associated HMG box protein (TOX) transcription factor, which is critically linked to T cell exhaustion [3, 5, 8]. Furthermore, we demonstrate that PD-1 blockade partially restores the cytotoxic, cytokine production, and proliferative capacity of these cells. This model very closely recapitulates many of the phenotypic and functional characteristics of patient-derived exhausted T cells and is ideal for the assessment of ICB treatments (as monotherapies or combinations).

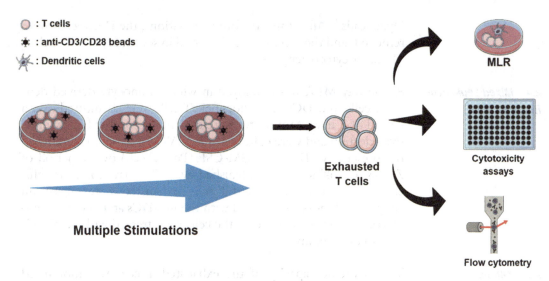

Fig. 1 In vitro generation of T_{EX} cells. The model is based on the repeated stimulation of isolated T cells with CD3/CD28 Dynabeads® over a period of 8 days. The in vitro-generated T_{EX} can be extensively characterized by flow cytometry and employed in secondary functional assays such as MLR or cytotoxicity assays

2 Materials and Methods

2.1 Cells and Cell Lines

Peripheral blood mononuclear cells (PBMCs) from healthy donors were employed to establish the assay. For the cytotoxicity assay, the ovarian cancer cell lines SKOV-3 and HeLa were employed. These cell lines were purchased by ATCC and subcultured according to provider's recommendations.

2.2 Antibodies

The following antibodies were used for cell staining: anti-CD3 PerCP/Cy5.5 (clone UHT1, BioLegend), anti-CD4 FITC (clone OKT4, BioLegend), anti-CD8 FITC (SK1, BioLegend), anti-PD-1 APC (clone MIH4, BioLegend), anti-LAG-3 PE (clone REA351, Miltenyi Biotec), anti-TIM-3 BV711 (clone 7D3, BD Biosciences), anti-CTLA-4 BV421 (clone BNI3, BioLegend), anti-CD137 BV650 (clone 4B4-1, BioLegend), and anti-TOX PE (clone REA473, Miltenyi Biotec).

2.3 Cell Culture

$CD4^+$, $CD8^+$, or pan-T cells were enriched from PBMCs using EasySep T cell isolation kits according to manufacturer's instructions. $1–3 \times 10^6$ T cells were plated in AIM-V medium (Life Technologies) supplemented with 1–5% heat-inactivated bovine serum (Sigma) in a 12- or 24-well plate. Cells were stimulated with T-Activator CD3/CD28 Dynabeads® (Life Technologies) following manufacturer's recommendations. Every 48 h, cells were counted, washed, and restimulated with a fresh batch of

Dynabeads®. After three or four stimulations, the Dynabeads were removed, and the T cells were employed in secondary assays such as MLRs or cytotoxicity assays.

2.4 Mixed Leukocyte Reaction (MLR)

A one-way MLR was employed in which monocyte-derived dendritic cells (mo-DCs) and allogeneic T cells were cocultured over a period of 5 days. Monocytes were enriched from PBMCs by positive selection and were cultured in X-VIVO 15 (Lonza) medium in the presence of IL-4 and GM-CSF (PeproTech) over a period of 7 days. Cytokines were replenished every 2–3 days to ensure efficient differentiation. Isolated T cells (exhausted or resting) from allogeneic donors were mixed with the mo-DCs at T:mo-DC ratio of 10:1. In several occasions, the cells were treated with pembrolizumab or nivolumab.

2.5 Kinetic Cytotoxicity Assay

The cytotoxic capacity of the exhausted T cells was monitored employing the xCELLigence Real-Time Cell Analyzer (RTCA) system (ACEA Biosciences). The xCELLigence RTCA instrument allows the target cells to be continuously monitored throughout the course of an experiment. Cell viability is determined through a noninvasive electrical impedance measurement. As cell death occurs, multiple biochemical changes take place that affect the size, shape, and number of cells that elicits a change of impedance, allowing the quantification of the phenomenon that directly relates to the magnitude of cell death that is occurring in the system.

SKOV-3 or HeLa cells were cultured with T cells (exhausted or resting) at different E:T ratios in the presence of CD3-/CD28-stimulating antibodies (clones UHT1 and CD28.2, BioLegend, respectively, at 2 μg/mL each). On occasions, the cells were treated with pembrolizumab, and T cell-mediated cytotoxicity was measured over a period of 4 days.

2.6 Flow Cytometry

Cultured T cells were firstly stained with fixable viability dye (eFluor 780, ThermoFisher) according to manufacturer's instructions and then with fluorescent conjugated antibodies for approximately 30 minutes in flow cytometry staining buffer (BioLegend). For nuclear staining, the eBioscience Foxp3/transcription factor staining buffer set was employed according to manufacturer's recommendations.

Flow cytometry of stained cells was performed using a Novocyte® Quanteon™ 4025 flow cytometer and data analyzed employing the NovoExpress® or FlowJo® software.

2.7 Statistics

Data was analyzed using GraphPad Prism 8. One-way or two-way ANOVA were performed according to experimental requirements.

3 Results

3.1 Repeated T Cell Stimulation Results in Sustained Expression of Exhaustion Markers

Based on the assumption that T cell exhaustion can be attributed to persistent antigenic stimulation, normal freshly isolated human T cells (pan-T cells) were continuously exposed to T cell receptor (TCR) and costimulatory stimulation with CD3/CD28 Dynabeads® over 2-day intervals. Repeated stimulation resulted in dramatic upregulation of the prototypical exhaustion markers PD-1, LAG-3, and TIM-3 relative unstimulated (resting T cells) or single-stimulated T cells (Fig. 2). More extensive characterization revealed that several other molecules linked to the exhaustion phenotype were upregulated in T cells exposed to repeated stimulation (relative to resting or single-stimulated T cells), including the checkpoint molecules CTLA-4 and BTLA and the ectonucleotidases CD39 and CD73 (Fig. 2). Beyond checkpoint molecules, these cells also expressed costimulatory molecules, namely, CD137 and CD134, confirming their activated status. Similar phenotypic characteristics were also observed when enriched $CD4^+$ or $CD8^+$ T cells were employed (data not shown). Recent studies have demonstrated that central to the molecular mechanism that induces and stabilizes the phenotypic and functional features of T cell exhaustion is the transcription factor TOX [3, 5, 8]. The expression of this transcription factor was assessed via flow cytometry on enriched T cells that had undergone the in vitro exhaustion protocol and compared to T cells that had undergone a single round of stimulation with CD3/CD28 Dynabeads®. Consistent with an exhausted phenotype, T cell that had undergone repeated stimulation expressed significantly higher levels of TOX relative to single-stimulated T cells (Fig. 3). This data clearly demonstrates that repeated T cell stimulation results in phenotypic changes in in vitro-generated T_{EX} cells that closely resemble patient-derived T_{EX} cells.

3.2 In Vitro-Generated T_{EX} Cells Are Functionally Impaired

As many of the checkpoint molecules are also upregulated during normal T cell activation, it was necessary to confirm the functional capabilities of the in vitro-generated T_{EX} cells. It has been shown previously that continuous stimulation of T cells with Dynabeads® results in the gradual downregulation of IL-2 and IFNγ production [4]. This was very prominent for IL-2, which initially was produced in high levels, followed by rapid loss of its production (Fig. 4). The loss of IL-2 production could not be attributed to increased IL-2 consumption as the secretion pattern was paralleled at the mRNA level [4]. A similar pattern was observed for IFNγ; however, the loss of production was not as dramatic as IL-2 (Fig. 4).

The exhausted T cells were then employed in an allogeneic one-way MLR. This system does not rely on antibodies for T cell stimulation and is a more physiological model of T cell activation.

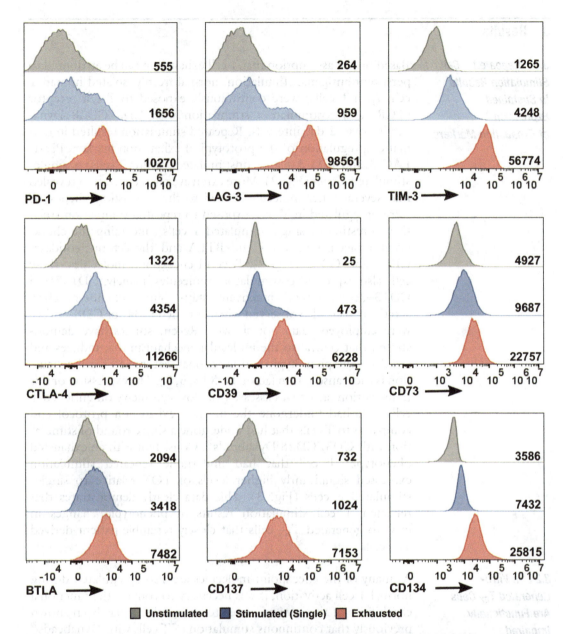

Fig. 2 In vitro-generated T_EX highly upregulate the expression of multiple checkpoint and costimulatory molecules. Isolated T cells (pan-T cells) were repeatedly stimulated with CD3/CD28 Dynabeads® over an 8-day period to generate T_EX cells, and the expression of PD-1, LAG-3, TIM-3, CTLA-4, CD39, CD73, BTLA, CD137, and CD134 was assessed by multiparameter flow cytometry (red histograms). As control, freshly isolated (unstimulated, gray histograms) or single-stimulated T cells (blue histograms) were employed. Example plots are gated on viable CD3+ T cells. Similar results were acquired in multiple donors (five) and in multiple experiments (minimally three times for each donor)

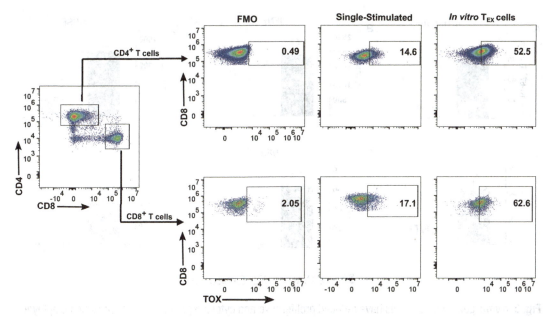

Fig. 3 In vitro-generated T_EX express high level of the transcription factor TOX. In vitro-generated T_EX or single-stimulated T cells were assessed for the expression of the transcription factor TOX by flow cytometry. T helper and cytotoxic T cells were identified by the expression of CD4 and CD8, respectively. Gates were set by employing fluorescence minus one (FMO) controls. Similar results were acquired in multiple donors and in multiple experiments

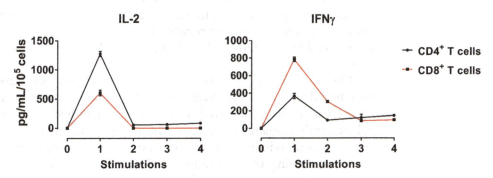

Fig. 4 Repeated CD3/CD28 stimulation results in reduced IL-2 and IFNγ production. Isolated CD4$^+$ or CD8$^+$ T cells were repeatedly stimulated with CD3/CD28 Dynabeads® (four times in total). After each stimulation, the cells were counted, supernatants were collected, and the expression of IL-2 (**a**) and IFNγ (**b**) was assessed by ELISA. Data depict mean ± SD ($n = 3$). Example data from one donor are presented. Similar data were acquired in multiple donors (five) and in multiple experiments (minimally three times for each donor)

The ability of the in vitro-generated T_EX cells to proliferate and produce cytokines was compared to freshly isolated resting T cells. Further confirming their functional impairment, the in vitro-generated T_EX cells demonstrated significantly reduced ability to proliferate in response to allogeneic T cell stimulation when

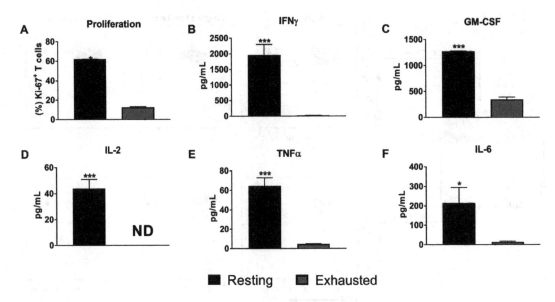

Fig. 5 In vitro-generated T_EX cells have reduced proliferative and cytokine production capacity when employed in an MLR. In vitro-generated T_EX (grey bars) or freshly isolated T cells (resting, black bars) were cocultured with allogeneic immature mo-DCs at a 10:1 ratio. After 5 days, the proliferative capacity of the T cells was assessed by flow cytometry by Ki-67 expression (a), while production of IFNγ, GM-CSF, IL-2, TNFα, and IL-6 was assessed by Luminex. Data is represented as mean ± SD ($n = 3$). $*p < 0.05$, $***p < 0.001$. Example data from one donor pair is presented. Similar data was acquired in multiple donor pairs (three) and in multiple experiments (minimally three times for each donor pair)

compared to resting T cells, as demonstrated by Ki-67 expression. Additional analysis demonstrated that these cells also demonstrated a diminished capacity (in comparison to resting T cells) to produce several cytokines, including IL-2, IFNγ, TNFα, IL-6, and GM-CSF (Fig. 5).

One of the most critical functions of T cells is to recognize and kill infected and transformed cells, and it has been demonstrated in several studies that exhausted T cells have diminished ability to induce cytolysis against these cells. The ability of the in vitro-generated T_EX cell to induce cytotoxicity of immortalized cell lines was investigated using the xCELLigence RTCA system. This instrument uses noninvasive electrical impedance to monitor cell proliferation, morphology change, and cell killing in a label-free, kinetic manner. The ability of the in vitro-generated T_EX, in comparison to freshly isolated or single-stimulated T cells, to induce cytotoxicity against the immortalized cancer cells lines, SKOV-3 and HeLa, was investigated. As there is no antigenic stimulus to redirect the T cells against the cancer cell lines, antibody-mediated CD3 and CD28 stimulation was employed for activation. Confirming their dysfunctional nature, the in vitro T_EX cells demonstrated a reduced ability to induce cytolysis of the target cell lines in comparison to the resting T cells (Fig. 6).

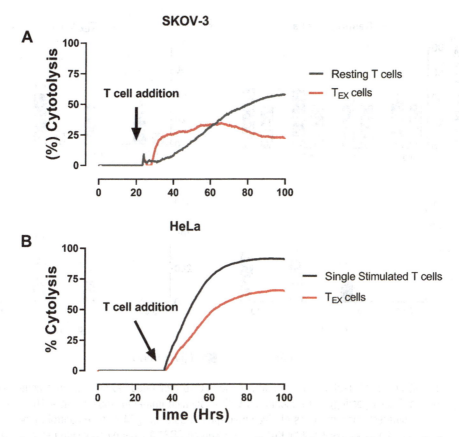

Fig. 6 In vitro-generated T$_{EX}$ have reduced cytotoxic capacity against immortalized cancer cells lines. (**a**) In vitro-generated T$_{EX}$ (red line) or freshly isolated T cells (resting, grey line) were cocultured with the ovarian cancer cell line SKOV-3 over a period of 4 days at 5:1 E:T ratio in the presence of CD3-/CD28-stimulating antibodies. Cytotoxicity was evaluated employing the xCELLigence RTCA MP instrument. (**b**) In vitro-generated T$_{EX}$ cells (red line) or single-stimulated T cells (grey line) were cocultured with the ovarian cancer cell line HeLa over a period of 4 days at an E:T ratio of 2:1 (in the presence of CD3-/CD28-stimulating antibodies), and cytotoxicity was evaluated with the xCELLigence RTCA MP instrument. Example data from one donor is presented. Similar data was acquired in multiple donors (three) and in multiple experiments

The above data clearly demonstrates that the in vitro T$_{EX}$ cell model captures the functional impairment of the patient-derived T$_{EX}$ cells.

3.3 PD-1 Inhibition Partially Restores Functionality of the In Vitro T$_{EX}$ T Cells

Targeting of the PD-1/PD-L1 pathway has revolutionized cancer treatment. Multiple studies have demonstrated that PD-1 has a central role in mediating the dysfunctional state that characterizes T cell exhaustion. Considering this, the ability of the PD-1-targeting antibodies nivolumab and pembrolizumab to restore the functionality of the in vitro-generated T$_{EX}$, in relation to their proliferative capacity and cytokine production levels, was assessed using an MLR system. Nivolumab was observed to have a small effect on proliferation when resting T cells were employed in the

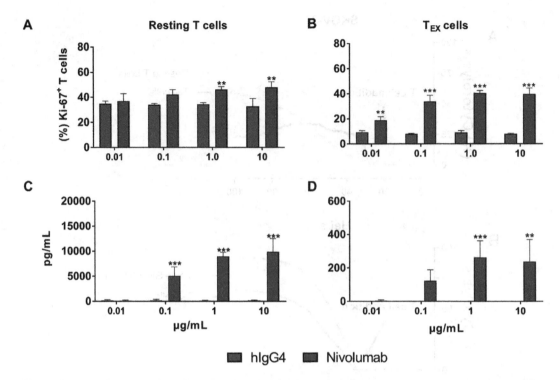

Fig. 7 PD-1 inhibition partially restores functionality of the in vitro-generated T_{EX} cells. In vitro-generated T_{EX} or freshly isolated T cells (resting) were cocultured with allogeneic immature mo-DCs at a 10:1 ratio and treated with three different concentrations of nivolumab (purple bars) or IgG4 isotype control (grey bars) for 5 days. (**a, b**) Proliferation was assessed by Ki-67 expression on CD3+ T cells by flow cytometry, (**c, d**) IFNγ production was assessed by ELISA. Example data from one donor pair is presented. Similar data was acquired in multiple donor pairs (three) and in multiple experiments (minimally three times for each donor pair). Data represent mean ± SD, $**p < 0.01$, $***p < 0.001$

MLR; however, it induced a dose-dependent enhancement of the proliferative capacity of the T_{EX} cells, which reached comparable levels to that of resting T cells (Fig. 7). In contrast to the proliferation readout, there was a clear dose-dependent effect of nivolumab on both resting and in vitro T_{EX} cells' ability to induce IFNγ production; however, the effect on the T_{EX} cells was modest, and even at the highest concentration of the antibody, only a ~5% of the IFNγ produced from resting T cells was observed (Fig. 7). Surprisingly, even though a clear effect could be observed on proliferation, nivolumab had no effect on IL-2 production, which is absent in exhausted T cells (data not shown).

The ability of PD-1 blockade to enhance the cytotoxic capacity of the in vitro T_{EX} cells was also investigated using the xCELLigence RTCA system and the ovarian cancer cell line SKOV-3. Pembrolizumab induced a dose-dependent increase of the cytotoxic potential of the T_{EX} cells, confirming the numerous published data for its immunostimulant activity; however, the cytotoxic activity of the T_{EX} cells collapsed during the later stage of the assay, and

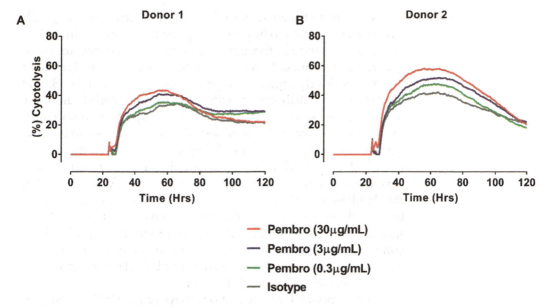

Fig. 8 PD-1 inhibition enhances the cytotoxic potential of the in vitro-generated T_{EX} cells. In vitro-generated T_{EX} cells were cocultured with the ovarian cancer cells SKOV-3 at a 5:1 E:T ratio and treated with three different concentrations of the PD-1-blocking antibody, pembrolizumab. Cytotoxicity was quantified employing the xCELLigence RTCA MP instrument over a period of 120 h. Example data from two donors is presented. Similar data was generated in further assessments

the effect of PD-1 diminished (Fig. 8). This transient effect of PD-1 blockade could be attributed to development of resistance to treatment as other regulatory pathways compensate; though, more studies are required to confirm this.

4 Discussion

Accumulating evidence in the last few years recognize the central role of T cell exhaustion in the pathophysiology of chronic viral infection and cancer. This has been further validated by the success of biologics targeting the PD-1/PD-L1 and CTLA-4/CD80-CD86 axis, which resulted in an acceleration of the efforts to understand the molecular pathways that regulate exhaustion with an aim to develop therapeutics that target them. However, efficient development of therapeutics requires robust human in vitro models that accurately emulate pathological conditions. The restimulation model, pioneered by Balkhi et al. [4] and further developed by us, fully captures the defining elements that distinguish exhausted T cells from differentiated effector and memory T cells. These include chronic stimulation, upregulation of checkpoint inhibitor molecules (PD-1, LAG-3, TIM-3), and loss of effector functions such as cytokine production, proliferative capacity, and cytotoxic failure.

Numerous characteristics of the in vitro-generated T_{EX} cells have been confirmed to be present in patient-derived T cells. Balkhi et al. demonstrated, through the application of the in vitro-generated T_{EX} model, that the transcription factor Yin Yang 1 (YY1) positively regulates the expression of PD-1, LAG-3, and TIM-3 and in collaboration with Ezh2 histone methyltransferase inhibits the production of IL-2 and IFNγ [4]. More importantly, they clinically validated this observation in melanoma-infiltrating lymphocytes and in PD-1$^+$ T cells in patients with HIV, confirming the physiological relevance of the model. Three seminal studies have demonstrated that the transcription factor TOX is central in the development of the T_{EX} phenotype in vivo and is expressed in tumor-infiltrating CD8$^+$ T cells and in human hepatitis C (HPC)-specific T cell clones from patients with chronic disease [3, 5, 8]. We confirmed the high expression of TOX in the in vitro-generated T_{EX} relatively to conventionally stimulated T cells, further validating the translational value of the assay.

Conventional T cell activation assays (e.g., MLR or antigen-specific T cell activation) can be informative and provide an excellent starting point for the assessment of molecules targeting checkpoint or costimulatory receptors. These assays may be misleading on the true biological potential of a molecule as they mostly employ fully functional T cells. This was clearly demonstrated when the effect of nivolumab was compared in MLRs that employed freshly isolated T cells (conventional MLR) or in vitro-generated T_{EX} cells. Nivolumab treatment had a very strong stimulating effect in IFNγ production in the conventional MLR; however, it had a much more modest (albeit consistent) effect on in vitro T_{EX} cells. On the other hand, the effect of proliferation was much more prominent in MLR reactions employing T_{EX} due to their lower proliferative capacity. Unlike freshly isolated T cells, the T_{EX} cells co-express multiple checkpoint and costimulatory molecules, which can explain the lack of complete reversal of exhaustion as other immunoregulatory pathways may compensate after PD-1 blockade. This also makes the in vitro-generated T_{EX} an ideal platform for developing assays for assessing combination therapies and bispecific (or multispecific) molecules.

In this chapter, we have demonstrated that the in vitro-generated T_{EX} cells are a versatile platform that can be employed to assess multiple cellular functions, including cytokine production, proliferation, and cytotoxic potential. We are currently working to further expand the potential of this model by employing methods such as single RNAseq that will allow identification of subpopulation of T_{EX} cells, transforming the assay into an even more powerful discovery and target validation tool.

References

1. Thommen DS, Schumacher TN (2018) T cell dysfunction in cancer. Cancer Cell 33:547–562. https://doi.org/10.1016/j.ccell.2018.03.012
2. Wherry EJ (2011) T cell exhaustion. Nat Immunol 12:492–499. https://doi.org/10.1038/ni.2035
3. Alfei F, Kanev K, Hofmann M et al (2019) TOX reinforces the phenotype and longevity of exhausted T cells in chronic viral infection. Nature 571:265–269. https://doi.org/10.1038/s41586-019-1326-9
4. Balkhi MY, Wittmann G, Xiong F, Junghans RP (2018) YY1 upregulates checkpoint receptors and downregulates type I cytokines in exhausted, chronically stimulated human T cells. iScience 2C:105–122. https://doi.org/10.1016/j.isci.2018.03.009
5. Khan O, Giles JR, McDonald S et al (2019) TOX transcriptionally and epigenetically programs CD8+ T cell exhaustion. Nature 571:211–218. https://doi.org/10.1038/s41586-019-1325-x
6. Mckinney EF, Lee JC, Jayne DRW et al (2015) T cell exhaustion, costimulation and clinical outcome in autoimmunity and infection. Nature 523:612–616. https://doi.org/10.1038/nature14468.T
7. Raghav SK, Deplancke B, Baitsch L et al (2011) Exhaustion of tumor-specific CD8+ T cells in metastases from melanoma patients. J Clin Invest 121:2350–2360. https://doi.org/10.1172/jci46102
8. Scott AC, Dündar F, Zumbo P et al (2019) TOX is a critical regulator of tumour-specific T cell differentiation. Nature 571:270–274. https://doi.org/10.1038/s41586-019-1324-y
9. Sen DR, Kaminski J, Barnitz RA et al (2016) The epigenetic landscape of T cell exhaustion. Science 354:1165–1169. https://doi.org/10.1126/science.aae0491
10. Trifari S, Hogan PG, Wong V et al (2017) Exhaustion-associated regulatory regions in CD8 + tumor-infiltrating T cells. Proc Natl Acad Sci 114:E2776–E2785. https://doi.org/10.1073/pnas.1620498114
11. Datar IJ, Sanmamed MF, Wang J et al (2019) Expression analysis and significance of PD-1, LAG-3 and TIM-3 in human non-small cell lung cancer using spatially-resolved and multiparametric single-cell analysis. Clin Cancer Res 4142:2018. https://doi.org/10.1158/1078-0432.CCR-18-4142
12. Anderson AE, Becker A, Yin F et al (2019) Abstract A124: preclinical characterization of AB154, a fully humanized anti-TIGIT antibody, for use in combination therapies. Cancer Immunol Res 7:A124. https://doi.org/10.1158/2326-6074.cricimteatiaacr18-a124
13. Du W, Yang M, Turner A et al (2017) Tim-3 as a target for cancer immunotherapy and mechanisms of action. Int J Mol Sci 18:1–12. https://doi.org/10.3390/ijms18030645
14. Friedlaender A, Addeo A, Banna G (2019) New emerging targets in cancer immunotherapy: the role of TIM3. ESMO Open 4:1–6. https://doi.org/10.1136/esmoopen-2019-000497

Chapter 7

Validation of an Image-Based 3D Natural Killer Cell-Mediated Cytotoxicity Assay

Brad Larson, Lubna Hussain, and Jenny Schroeder

Abstract

An in vitro model for natural killer (NK) cell-mediated cytotoxicity was developed using a collagen-based scaffold 3D cell culture technology. This technology induced HCT116 cells to aggregate into tumoroids over time, which became the target cells during the cytotoxicity assay. Cytotoxicity was assessed by both phosphatidyl serine exposure (apoptosis) and plasma membrane rupture (necrosis) using fully automated workflows. Cytotoxicity was quantified using NK cells alone and with IL-2 stimulation, where a significant increase of cytotoxicity was evident. Cytotoxicity with this model was compared to HCT116 cells adhered to microplates in a conventional two-dimensional (2D) format. The three-dimensional (3D) model was far superior in both maintaining cell health over time and accurately depicting cytotoxic events.

Key words Tumoroid, HCT116, Natural killer cell, Cell-mediated cytotoxicity, Live cell imaging

1 Introduction

Natural killer (NK) cells are cytotoxic lymphocytes found in peripheral blood that play a role in host defense and immune regulation. NK cells are particularly interesting in immunotherapy due to their potential to target and destroy specific cancer cells, while leaving non-target healthy cells intact. The anticancer activity of NK cells is shown to be associated with an improved prognosis in several cancers such as colorectal cancer [1], non-small cell lung cancer [2], and clear cell renal cell carcinoma [3].

To properly study the interaction between NK cells and target tumor cells, an appropriate in vitro model system must be established. However, much of the data published to date used cancer cells plated as a two-dimensional (2D) monolayer on the bottom of microplate wells. A growing amount of data has shown that cells cultured in this manner lack the cell–cell and cell–matrix communication, metabolic gradients, and polarity demonstrated in vivo [4]. The ability to perform matrix infiltration studies is also eliminated with the use of 2D cell culture. By embedding cancer cells into a three-dimensional (3D) matrix and allowing the formation of tumoroids, the shortcomings of using 2D cultured cells can be overcome as communication networks and cellular gradients observed within in vivo tumors are reestablished.

With the incorporation of 3D cultured cells, however, traditional methods to monitor target and NK cell interactions, and subsequent target cell killing can become problematic. Microplate reader assays designed to detect signal from cell monolayers lack the sensitivity to quantify signal from tumoroids surrounded by non-cell containing areas in the well with no signal generation. By incorporating microscopy-based detection and cellular analysis, signal emanating solely from tumoroids is quantified, providing a highly robust method to detect induced toxicity within target cancer cells.

Here, we describe a novel 3D NK cell-mediated cytotoxicity (CMC) assay. Using the RAFT™ 3D Cell Culture System principle (Fig. 1), HCT116 colorectal cancer cells were embedded within a collagen hydrogel of defined concentration and thickness, mimicking in vivo extracellular matrix (ECM). Following cell propagation to create tumoroids within the matrix, HCT116 and NK cells were labeled with individual cell tracking dyes, followed by NK cell addition. Fluorescent apoptosis and necrosis probes were also added to track cytotoxic events within the tumoroids. Cellular imaging and analysis were performed at regular intervals over a 7-day period to monitor NK cell binding to the tumoroids and induced apoptosis and necrosis of the HCT116 cells making up each tumoroid. Experimental testing validated that the combined assay technique provides a sensitive, accurate, and repeatable in vitro method to determine the ability of NK cells to target and kill tumor cells.

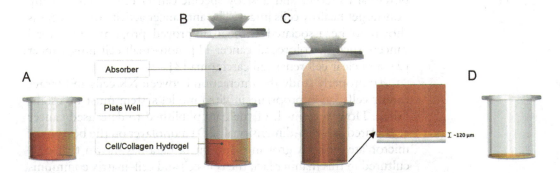

Fig. 1 Creation of 3D cell/collagen hydrogel using RAFT™ System. (**a**) Cell–collagen mix dispensed to wells of tissue culture (TC)-treated plate. (**b**) Absorber insertion into plate well. (**c**) Absorption of medium, concentrating collagen and cells to in vivo strength, creating an ~120 μm thick hydrogel. (**d**) Removal of absorber prior to dispense of fresh cell medium

2 Materials and Methods

2.1 Materials

2.1.1 Cells

HCT116 epithelial colorectal carcinoma cells (Catalog No. CCL-247) were obtained from ATCC (Manassas, VA). Negatively selected peripheral blood CD56+ CD16+ natural killer (NK) cells (Catalog No. 2W-501) were donated by Lonza (Walkersville, MD).

2.1.2 Assay and Experimental Components

The RAFT™ 96-well Small Kit, including reagents and absorbers, (Catalog No. 016-1R17) and LGM™-3 (Lymphocyte Growth Medium-3) (Catalog No. CC-3211) were generously donated by Lonza. 96-well TC-treated microplates (Catalog No. 655090) were donated by Greiner Bio-One, Inc., (Monroe, NC). CellTracker™ Deep Red dye (Catalog No. C34565) was procured from ThermoFisher Scientific (Waltham, MA). Kinetic Apoptosis Kit, containing pSIVA™-IANBD and propidium iodide reagents (Catalog No. ab129817) was donated by abcam (Cambridge, MA), and IL-2 IS (Catalog No. 130-097-744) was donated by Miltenyi Biotec (San Diego, CA).

2.1.3 Cytation™ 5 Cell Imaging Multimode Reader

Cytation™ 5 is a modular multimode microplate reader combined with automated digital microscopy. Filter- and monochromator-based microplate reading are available, and the microscopy module provides up to 60× magnification in fluorescence, brightfield, color brightfield, and phase contrast. The instrument can perform fluorescence imaging in up to four channels in a single step. With special emphasis on live-cell assays, Cytation™ 5 features temperature control to 65 °C, CO_2/O_2 gas control and dual injectors for kinetic assays, and is controlled by integrated Gen5™ Microplate Reader and Imager Software. The imager and software were used to capture brightfield and fluorescent images for 2D and 3D CMC assays and quantify the level of NK induced cytotoxicity.

2.1.4 BioSpa™ 8 Automated Incubator

The BioSpa™ 8 Automated Incubator links BioTek readers or imagers together with washers and dispensers for full workflow automation of up to eight microplates. Temperature, CO_2/O_2, and humidity levels are controlled and monitored through the BioSpa™ software to maintain an ideal environment for cell cultures during all experimental stages. Test plates were incubated in the BioSpa™ and automatically transferred to the Cytation™ 5 at designated time points to monitor cytotoxicity in 2D and 3D cultured cells.

2.2 Methods

2.2.1 2D and 3D Target and NK Effector Cell Preparation

On Day 0, HCT116 target cells were mixed with the prepared RAFT™ Collagen I Suspension and dispensed to 96-well tissue culture (TC)-treated microplates in a volume of 240 μL to yield 1200 cells/well. The cell plate was incubated at 37 °C/5% CO_2 for 15 min, followed by addition of the absorbers in the RAFT™ Plate, and a second 15-min incubation at 37 °C/5% CO_2 during which the RAFT™ Process concentrated the collagen density to a physiologically relevant strength. The absorbers were then removed and 100 μL of new medium was then added to the concentrated cell/collagen hydrogel (Fig. 1). Cells were propagated at 37 °C/5% CO_2 in the plates for 7 days to allow tumoroid creation through cell doubling, with media exchanges every 2 days. On Day 6, a total of 3000 cells/well of HCT116 cells intended for 2D culture were added to a separate 96-well TC-treated microplate and allowed to attach overnight.

On Day 7, negatively selected peripheral blood CD56+ CD16+ natural killer (NK) cells were prepared. Cryopreserved NK cells were thawed and diluted in LGM™-3 (Lymphocyte Growth Medium-3) per the manufacturer's protocol, then stained with the fluorescent CellTracker™ Deep Red dye. The NK cells were diluted to a concentration that equaled a 20:1 or 10:1 ratio of final HCT116 populations per well in media containing either Abcam Kinetic Apoptosis Kit pSIVA™-IANBD and propidium iodide reagents and 500 U/mL human IL-2 IS, or apoptosis kit reagents alone, and added to the 3D tumoroids and the 2D cultured cells.

2.2.2 Automated Cell-Mediated Cytotoxicity Process

After the diluted NK cell mixtures were added to the 96-well 3D and 2D test plates, the plates were placed into the BioSpa™ 8 at 37 °C/5% CO_2 for an incubation period of 120 h. The BioSpa™ method was programmed such that plates were automatically moved to the Cytation™ 5 every 2 h, where fluorescent imaging was carried out to monitor NK effector cell induced cytotoxicity (Fig. 2). The Cytation™ 5 imaging chamber was also maintained at

Fig. 2 Automated cell-mediated cytotoxicity system consisting of BioSpa™ 8 Automated Incubator (left) and Cytation™ 5 Cell Imaging Multimode Reader (right)

37 °C/5% CO_2 to ensure consistent environmental cell conditions. Imaging channels included in the experimental procedure were as follows: GFP: pSIVA™-IANBD fluorescent probe binding to externally exposed phosphatidyl serine (PS) on apoptotic cells; propidium iodide (PI): PI intercalating dye bound to necrotic cell DNA; CY5: CellTracker™ Deep Red stained NK cells.

3 Results and Discussion

3.1 Tumoroid Formation Conformational Imaging

Brightfield imaging was performed to confirm that 3D tumoroids formed over the 7-day incubation period via HCT116 propagation.

As seen by the images in Fig. 3, HCT116 cells initially suspended in the RAFT™ Hydrogel Matrix are able to continuously divide over the 7 day incubation period. 3D growth and tumoroid assembly is also confirmed by the images captured at separate Z-Heights in Fig. 3a–c. A final in-focus image is able to be created using the Z-projection capabilities in the Gen5™ software (Fig. 3d).

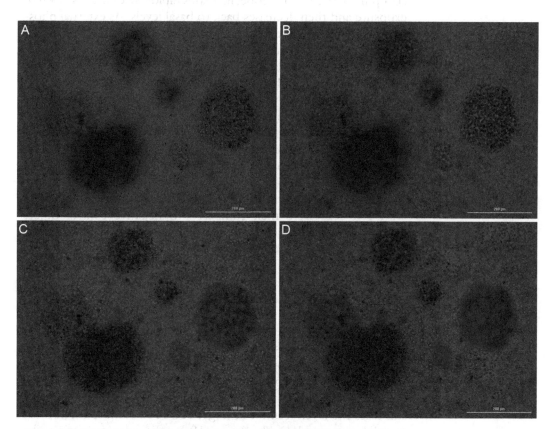

Fig. 3 3D Tumoroid formation conformational imaging. (**a–c**) Brightfield image captured using 10× objective at three separate Z-heights within the RAFT™ Hydrogel. (**d**) Final Z-projected image of tumoroids

3.2 3D NK Cell-Mediated Cytotoxicity Imaging

Following NK cell addition to the 3D plates, Z-stacked images were automatically captured of HCT116 tumoroids every 2 h over the entire 120 h incubation period from 20:1 to 10:1 NK-treated positive control wells, as well as negative control wells containing no NK cells (Fig. 4).

By using overlaid final projections of the Z-stacked images, NK cell interactions with HCT116 tumoroids, in addition to apoptotic and necrotic cell induction, were able to be accurately tracked in a kinetic fashion for each test condition.

3.3 3D NK Cell Apoptosis and Necrosis Induction Analysis

Using the Z-stacked images, Gen5™ software automatically pre-processed the samples to remove excess background signal and prepare the image for quantitative analysis. Two separate cellular analysis steps were conducted to place object masks around areas within the image, meeting primary analysis criteria in either the GFP or PI imaging channel. Minimum and maximum object sizes and threshold fluorescence values were set such that only apoptotic or necrotic areas within target tumoroids were identified (Fig. 5a, b).

The graphs in Fig. 5c, d demonstrate how the cell area within tumoroids treated with NK cells, identified with the green fluorescent pSIVA™-IANBD probe, increases rapidly over the first 24 h of exposure and then decreases back to basal levels. As external phosphatidyl serine (PS) exposure is a hallmark of early apoptotic activity, and the reagent binds to externally exposed PS, the observed phenomenon is consistent with expected results as apoptosis is initially induced within treated tumoroid cells. As NK cells continue to interact with tumoroids, cells become increasingly necrotic and PS once again internalizes within the lipid bilayer. This is confirmed by the increase in signal from the fluorescent cell impermeable propidium iodide (PI) probe (Fig. 5e, f). Cellular necrosis leads to loss of membrane integrity, allowing PI to bind to the nucleus.

Object area values from wells containing tumoroids treated with NK cells were then compared to negative control object areas at individual time points to determine the level of apoptotic or necrotic activity within tumoroids induced by stimulated or unstimulated NK and HCT116 target cell interaction (Fig. 6a–d).

Due to the fact that untreated HCT116 cells, aggregated into 3D tumoroids, maintain high cell viability, and negligible observable cytotoxicity, apoptotic and necrotic fold increase graphs follow a similar pattern as the raw coverage area graphs plotted in Fig. 5. This further validates that exposure to NK cells causes initial HCT116 cellular apoptosis followed by secondary necrosis, and also agrees with previously published literature results [5].

From the data, it is also apparent that the 20:1 NK cell ratio increases the level of induced target HCT116 tumoroid cell cytotoxicity compared to the lower 10:1 ratio, demonstrating the cumulative effect of NK cells on cancer tumoroids. Finally,

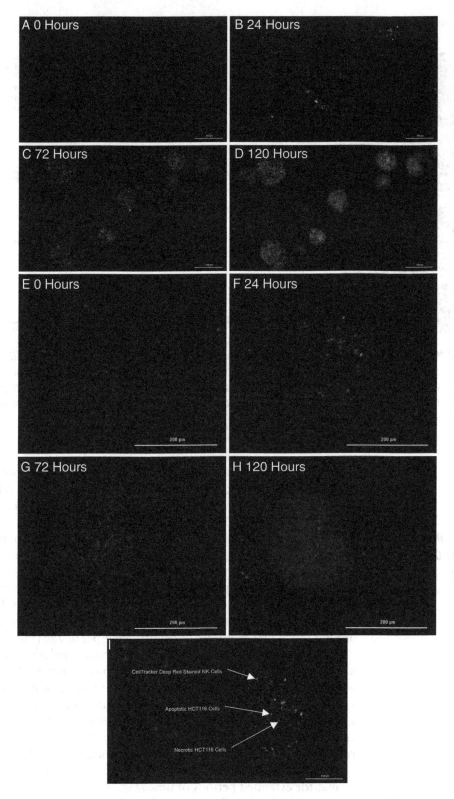

Fig. 4 Imaging of NK cell induced apoptosis and necrosis within HCT116 tumoroids. (**a–d**) Fluorescent overlaid final projected multi tumoroid images captured using a 4× objective. (**e–h**) Zoomed single tumoroid images following treatment with a 20:1 ratio of IL-2 stimulated NK cells and 0, 24, 72, or 120 h incubations, respectively. (**i**) Projected, zoomed image demonstrating green pSIVA™-IANDB stained apoptotic cells, orange propidium iodide stained necrotic cells, and red CellTracker™ Deep Red labeled NK cells

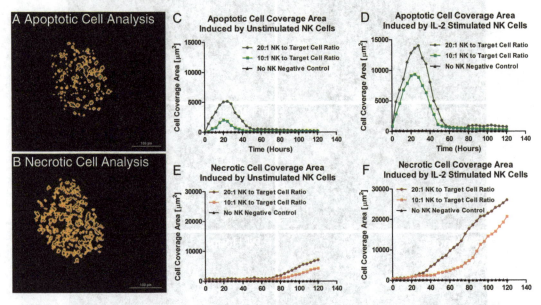

Fig. 5 3D NK CMC analysis. Gen5™ placed object masks around (**a**) apoptotic or (**b**) necrotic cells during individual analysis of images captured from 3D test plates. Total cell coverage area calculated from object masks placed around all apoptotic cells within tumoroids following interaction with (**c**) unstimulated or (**d**) IL-2 stimulated NK cells; and necrotic cell coverage area following interaction with (**e**) unstimulated or (**f**) IL-2 stimulated NK cells at each time point

stimulation by IL-2 was also shown to positively influence NK induced cytotoxicity compared to resting inactivated NK cells at both ratios, again agreeing with previous literature findings [6].

3.4 2D NK Cell-Mediated Cytotoxicity Imaging

Kinetic montage images were captured from test plates containing 2D cultured HCT116 and NK cells or HCT116 cells alone using the aforementioned automated monitoring procedure (Fig. 7).

Following image pre-processing and stitching, individual primary analyses were once again carried out using either the GFP or PI on the final complete images. Here Gen5™ placed object masks around individual apoptotic or necrotic cells per image (Fig. 8a, b).

NK cell induction of apoptotic and necrotic activity follows the same basic pattern for 2D cultured HCT116 cells (Fig. 8c–f) as that seen for 3D cultured cells (Fig. 5c, d) up to 80 h of incubation. Greater NK cell ratios, in addition to IL-2 stimulation, again exert a greater toxic effect. However, after 80 h the number of identified apoptotic and necrotic untreated cells begins to increase. Upon review of captured images from negative control wells, this is due to decreased cell health when HCT116 cells are cultured in 2D format for extended periods of time (Fig. 9).

Due to the fact that untreated cells become apoptotic and necrotic, final fold stimulation values can be affected. The affect is negated when examining apoptotic induction, as cells initially

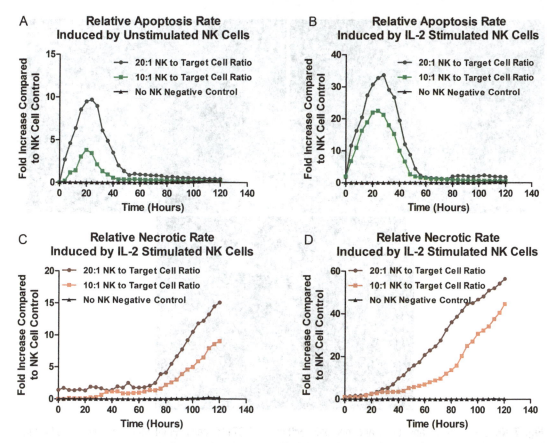

Fig. 6 3D Cytotoxicity fold induction calculation. Fold stimulation calculated of induced apoptotic activity from interaction with (**a**) unstimulated or (**b**) IL-2 stimulated NK cells; and induced necrotic activity from interaction with (**c**) unstimulated or (**d**) IL-2 stimulated NK cells at each time point

induced to undergo apoptosis become necrotic at approximately the same time as uninduced cells naturally become apoptotic from time spent in the well. Therefore, the graphs seen in Fig. 10a, b exhibit the same curve shape as those seen in Fig. 6a, b. A more dramatic change is seen when examining necrotic activity fold stimulation graph. As cell necrosis was negligible in 3D cultured HCT116 cells, fold stimulation values continued to increase over time. However, due to increasing numbers of necrotic cells in untreated wells, fold stimulation values from NK cell exposure actually begin to decrease after a rapid initial increase.

The observed loss of cell viability in untreated 2D cultured HCT116 cells demonstrates the limitations of using target cells cultured in this manner to determine the long-term effect that NK cell based immunotherapy may have on cancer tumors. This limitation can be overcome by culturing these same cells in a 3D configuration.

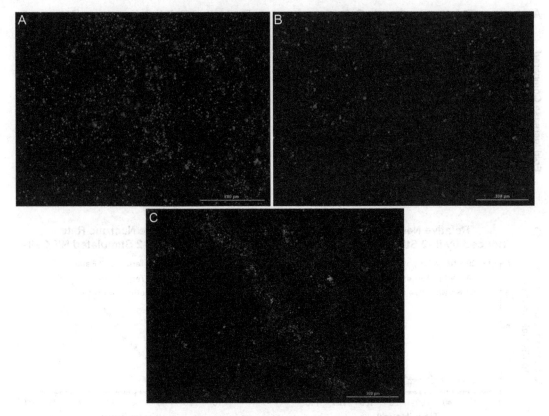

Fig. 7 NK cell induced apoptosis and necrosis within 2D HCT116 cells. Final stitched images following treatment with a 20:1 ratio of IL-2 stimulated NK to 2D cultured HCT116 cells following (**a**) 0, (**b**) 24, or (**c**) 120 h incubations. Green: pSIVA™-IANBD stained apoptotic cells; yellow: propidium iodide stained necrotic cells; red: CellTracker™ Deep Red labeled NK cells

4 Conclusions

The RAFT™ 3D Cell Culture System provides an ideal method to prepare and study co-cultures of tumoroids and immune cells. Furthermore, peripheral blood CD56+ CD16+ natural killer cells can be used with 3D and 2D cultured target cells to assess potential CMC activity. By stimulating NK cells with IL-2 the cytotoxic effect of NK cells on 3D and 2D cultured target HCT116 cells can be increased. Monitoring of apoptotic and necrotic cell induction can be carried out in an automated image-based manner, while not sacrificing proper cellular incubation conditions, through incorporation of the BioSpa™ 8 and Cytation™ 5. With the inclusion of target cancer cells cultured as 3D tumoroids, long-term CMC assays can be performed while maintaining negligible levels of uninduced basal cytotoxicity in target cells. Finally, the

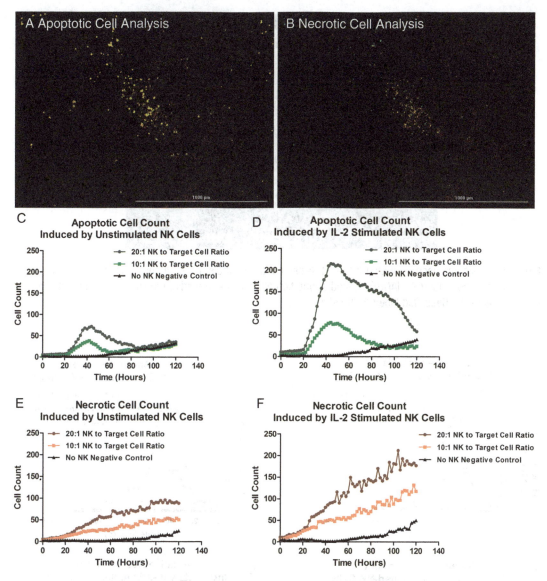

Fig. 8 2D NK CMC analysis. Gen5™ placed object masks around (**a**) apoptotic or (**b**) necrotic cells during individual analysis of images captured from 3D test plates. Total cell numbers calculated from object masks placed around all apoptotic cells per image following interaction with (**c**) unstimulated or (**d**) IL-2 stimulated NK cells; and necrotic cells following interaction with (**e**) unstimulated or (**f**) IL-2 stimulated NK cells at each time point

combination of appropriate 3D cell models, assay methodology, and walk away automation provide a robust process to generate accurate in vitro NK cell-mediated cytotoxicity results.

Fig. 9 Apoptosis and necrosis within 2D HCT116 cell culture controls. Final stitched images following 120 h incubations. Green: pSIVA™-IANBD stained apoptotic cells; yellow: propidium iodide stained necrotic cells; red: CellTracker™ Deep Red labeled NK cells

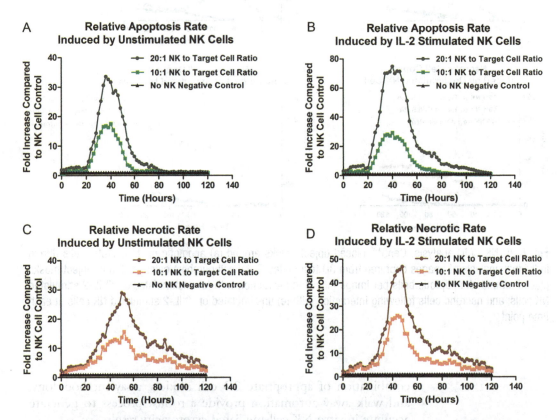

Fig. 10 2D cytotoxicity fold induction calculation. Fold stimulation calculated of induced apoptotic activity from interaction with (**a**) unstimulated or (**b**) IL-2 stimulated NK cells; and induced necrotic activity from interaction with (**c**) unstimulated or (**d**) IL-2 stimulated NK cells at each time point

References

1. Ohtani H (2007) Focus on TILs: prognostic significance of tumor infiltrating lymphocytes in human colorectal cancer. Cancer Immun 7(4):1–9
2. Bruno A, Ferlazzo G, Albini A, Noonan DM (2014) A think tank of TINK/TANKs: tumor-infiltrating/tumor-associated natural killer cells in tumor progression and angiogenesis. J Natl Cancer Inst 106(8):1–13
3. Eckl J, Buchner A, Prinz PU, Riesenberg R, Siegert SI, Kammerer R, Nelson PJ, Noessner E (2012) Transcript signature predicts tissue NK cell content and defines renal cell carcinoma subgroups independent of TNM staging. J Mol Med (Berl) 90(1):55–66
4. Hirschhaeuser F, Menne H, Dittfeld C, West J, Mueller-Klieser W, Kunz-Schughart LA (2010) Multicellular tumor tumoroids: an underestimated tool is catching up again. J Biotechnol 148(1):3–15
5. Blom WM, De Bont HJ, Meijerman I, Kuppen PJ, Mulder GJ, Nagelkerke JF (1999) Interleukin-2-activated natural killer cells can induce both apoptosis and necrosis in rat hepatocytes. Hepatology 29(3):785–792
6. Lehmann C, Zeis M, Uharek L (2001) Activation of natural killer cells with interleukin 2 (IL-2) and IL-12 increases perforin binding and subsequent lysis of tumour cells. Br J Haematol 114(3):660–665

Chapter 8

3D-3-Culture: Tumor Models to Study Heterotypic Interactions in the Tumor Microenvironment

Sofia P. Rebelo, Catarina Pinto, Nuno Lopes, Tatiana R. Martins, Paula Marques Alves, and Catarina Brito

Abstract

The potential of immunotherapies for the treatment of cancer has been shown in the success of checkpoint blockade therapies and adoptive T cell transfer approaches. These can mediate complete tumor regression in a subset of patients; however, not all patients respond to the treatment. One of the working hypotheses emphasizes the role of the tumor microenvironment (TME) surrounding the lesion, which can result in the inefficacy of immunotherapy in most cases. Different strategies are emerging to tackle the immunosuppressive nature of the TME, and more accurate preclinical models are needed for accurate evaluation of those as a majority of these models present compromised immune system and offers non-human tumor-stromal interactions. Therefore, there is an urgent need to incorporate immune cells within preclinical models, so that these questions can begin to be addressed in a human relevant setting and that therapeutic biomarkers for effective patient stratification are identified. In this chapter, we provide a detailed description of the procedures for implementing and using the culture system described previously by us (Rebelo et al., Biomaterials 163:185–197, 2018). This consists of a 3D cell model, the 3D-3-culture, enclosing three cellular components: tumour spheroids, cancer-associated fibroblasts, and monocytes. The model is based on alginate microencapsulation, which allows direct interaction between different cell types and is compatible with continuous monitoring and functional assessment in high-throughput and high-content screening.

Key words 3D culture, Spheroid, Macrophage, Myeloid, Immune cell infiltration, Bioreactor, Alginate, Tumor, In vitro, Tumor microenvironment

Abbreviations

CAF	Cancer-associated fibroblasts
CDX	Cell-derived xenografts
CSF1R	Colony-stimulating factor 1 receptor
DPBS	Dulbecco's phosphate-buffered saline
ECM	Extracellular matrix
EDTA	Ethylenediaminetetraacetic acid
FBS	Fetal bovine serum
FDA	Fluorescein diacetate

GEMM	Genetically engineered murine models
GFP	Green fluorescence protein
HEPES	4-(2-Hydroxyethyl)-1-piperazineethanesulfonic acid
NSCLC	Non-small cell lung carcinoma
PBMC	Peripheral blood mononuclear cells
PBS	Phosphate-buffered saline
PD1	Programmed cell death protein 1
PDL1	Programmed cell death-ligand 1
PDX	Patient-derived xenografts
rpm	Rotations per minute
RPMI	Roswell Park Memorial Institute medium
RT	Room temperature
TAM	Tumor associated macrophages
TME	Tumor microenvironment

1 Introduction

Cancer research has increased considerably in the last decades. Concomitantly, several therapeutic breakthroughs have arisen, resulting in overall treatment improvement and higher patient survival rate. Nevertheless, when considering the efforts to develop novel therapeutics, it is evident that the drug development in oncology is still an inefficient and costly process, mostly due to high attrition rates. Indeed, oncology is the disease area where investigational drugs have the lowest likelihood of reaching market approval after entering phase I (only 6.7%) [1]. One of the critical factors that can contribute to the reduction of the attrition rates is the use of better pre-clinical models that can capture tumor complexity and predict patient response.

Animal models such as genetically engineered murine models (GEMM) and cell or patient-derived xenografts (CDX/PDX) allow studying gene and protein functions in vivo, and contribute to a better understanding of their molecular pathways and underlying mechanisms. Nevertheless, the species-specific differences and/or the lack of a human-relevant immune microenvironment limit their predictive value and drive the need for the use of human systems in pre-clinical development [2]. In addition, it is critical that these models are reproducible, scalable, and compatible with high-throughput & high-content platforms, so that they can be easily integrated in screening campaigns. Most in vitro models used so far are based on cell lines expanded as monolayers, due to their easy set-up and technical feasibility. Nevertheless, the cell morphology and polarity are often altered in these systems. Three-dimensional

---

Sofia P. Rebelo and Catarina Pinto contributed equally to this work.

models such as tumor spheroids overcome some of these limitations, as there is maximization of cell-cell and cell-matrix interactions which results in extracellular matrix (ECM) accumulation and formation of tissue-like gradients of nutrients, metabolites, and oxygen [3–5]. Therefore, tumor spheroids have been applied to study tumor signalling and have gained spotlight in translational studies. In addition, 3D cultures have lower immunogenicity when compared to monolayers [6, 7], highlighting its relevance to study the dynamics between immune and tumor cells. There are several methods to generate tumor spheroids, such as low-adhesion plates, induction of aggregation in stirred-tank systems or embedment in bioactive scaffolds (e.g., matrigel or collagen) [8]. Aggregation in stirred-tank systems is a scalable approach, which generates a large number of spheroids with high homogeneity [9]. Bioactive scaffolds may have batch-to-bach variation and the undefined components may also alter the phenotype of the cells [10, 11], which can modulate cell behavior and impact assay reproducibility. Alginate is a polysaccharide hydrogel that is inert and biocompatible, being used as scaffold for cell entrapment [12]. It is composed of 99% water, but retains plasticity and mechanical strength. Gelling occurs by crosslinking with divalent ions, allowing cell entrapment under physiological conditions and cell recovery by gel dissolution, enabling cell manipulation and characterization after culture.

In the recent years, the knowledge on the bidirectional interaction between tumor cells and their TME has brought novel perspectives on the therapeutic approach to cancer. Cancer associated fibroblasts (CAF) are major players of the TME. Upon modulation by tumor cells, CAF augment secretion of ECM components (namely collagen and fibronectin) and inflammation factors, which results in tissue stiffness and contributes to tumor progression [13]. The interaction between tumor and immune cells also plays a critical role in all steps of tumor development. Immune infiltrates are composed of both innate (macrophages, natural killer, mast cells) and adaptive (T and B lymphocytes) immune cells which stop to perform their role of controlling tumor growth after contacting with the TME [14]. Macrophages, part of the innate immune system, may represent up to 50% of the tumor mass in some tumor indications [15, 16]. These are recruited to the tumor tissue from circulating monocytes and are modulated by the soluble factors and cell-cell and cell-ECM interactions within the TME. Tumor associated macrophages (TAM) exert tumor-promoting effects such as induction of proliferation, angionesis, ECM remodeling, and evasion from adaptive immunity, characteristics typically associated with the M2-like immunosuppressive phenotype.

Nowadays, however, in light of the extensive network of transcriptional regulators at play during macrophage polarization, characterization of macrophage populations is based on a spectrum of phenotypes, which need further characterization [17]. Also, there have been additional macrophage molecular phenotypes identified that do not reflect the current M1/M2 paradigm [18]. The first single-cell analysis of the innate immune compartment in human cancer patients evidenced distinct myeloid cell subsets and their differential contribution to anti-tumor T cell immunity [19]. Understanding the changes occurring within these subsets in the initial tumor lesion can provide immunomodulatory strategies targeting the innate immune compartment [19].

In fact, promising results in both animal models and clinical trials targeting TAM suggests an important role in combination therapies [20]. Macrophages cross-present foreign antigens and produce T cell differentiation cytokines, playing an important role in managing T cell function at the tumor site [19]. Therefore, controlling the macrophage compartment or changing macrophage phenotype at the tumor site may serve as a synergistic role in potentiating T cell targeting therapies [19]. Indeed, decreased macrophage infiltration leads to higher efficacy of immunotherapies [21]. Interestingly, PD1/PDL1 blockade has also been shown to restore TAM phagocytic capacity and reduce tumor growth of colorectal cancer in mice models [22].

Thus, the TME not only influences therapeutic response but also it is now regarded as an interesting therapeutic target. To better understand the TME response to standard of care therapies, its influence on tumor cells, and to evaluate novel TME-targeting drugs, it is vital to integrate TME components into more complex tumor cell models. This chapter describes a strategy for the establishment of complex tumor models integrating tumor, stromal, and immune cells, based on previous work established in our laboratory [12, 23]. Herein, is presented a protocol for aggregation of tumor cells in stirred systems and for the coculture of microencapsulated tumor spheroids with stromal and myeloid cells (Fig. 1). The coculture model, based on the alginate microencapsulation strategy, allows direct interaction between the different cell types and is compatible with continuous monitoring and functional assessment in long-term culture using stirred systems. Additionally, the alginate microcapsules are permeable to small compounds and thus can be used for drug testing. As proof-of-concept, the effects of two standard-of-care chemotherapeutic drugs—paclitaxel and cisplatin—were tested. The results were in line with those reported in the literature for the referred compounds. Moreover, the plasticity and high response to stimuli characteristic of TAM was also supported in the developed model. Upon challenge with an immunomodulatory agent—a CSF1R inhibitor—the macrophage phenotype was modulated towards a more pro-inflammatory phenotype [23]. Overall, these results suggested that models based on this culture system can be incorporated into drug development pipelines, including immunotherapeutics.

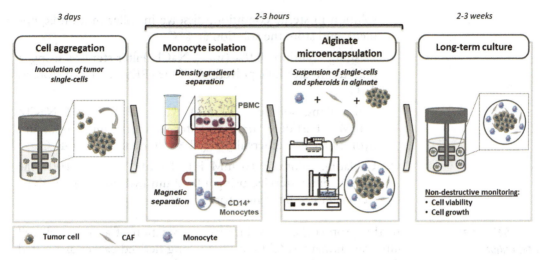

Fig. 1 Workflow for preparation of the microencapsulated 3D-3-coculture cell model. Briefly, tumor cells are inoculated as single cells in stirred-tank vessels to promote cell aggregation. After 3 days, the monocyte isolation procedure is initiated. Peripheral blood mononuclear cells (PBMC) are obtained by density gradient centrifugation. By magnetic separation, CD14-positive monocytes are selected. The monocytes and CAF single-cell suspension is mixed with the tumor spheroids in alginate solution and the microencapsulation procedure is initiated. After the coculture encapsulation within the alginate beads, the culture is maintained in stirred culture for the long term. The distinct cell types can be recovered by dissolving the alginate microcapsules with a chelating solution. The cells can be used for characterization or other functional studies. The image is not at scale

2 Materials

2.1 General Reagents and Equipment

1. Biological safety cabinet (biosafety level II).
2. Inverted bright-field and fluorescence microscope.
3. Humidified incubator with CO_2 control.
4. Flow cytometer.

2.2 Microencapsulation Equipment and Reagents

1. Electrostatically driven microencapsulation unit (VarV1 from Nisco), including:
   (a) Nozzle (Ø 0.17/0.7 mm).
   (b) Nozzle holder with autoclavable arm.
   (c) Electrode.
   (d) Magnetic agitator 0.
2. Syringe pump (KDS100).
3. Disposable material: Petri dishes, silicon tubes, and syringes (Terumo, 5 mL).
4. Autoclavable material: tweezers and aluminum box.
5. Alginate solution: 1.1% (w/v) Alginate Pronova UP MVG (Novamatrix) in 0.9% (w/v) of NaCl. Prepare the alginate

solution in sterile condition, dissolve in roller motion equipment, and store the solution at 4 °C.

6. Washing solution: 0.9% (w/v) NaCl. Filter-sterilize using a 0.22 μm hydrophilic polyethersulfone (PES) filter and store at 4 °C.

7. Crosslinking solution: 20 mM $BaCl_2$ in 115 mM NaCl/ 5 mM L-Histidine pH 7.4. Filter-sterilize using a 0.22 μm hydrophilic polyethersulfone (PES) filter and store at 4 °C.

8. Chelating solution: 10 mM HEPES pH 7.4 with 100 mM EDTA. Filter-sterilize using a 0.22 μm hydrophilic polyethersulfone (PES) filter and store at 4 °C.

2.3 Cell Isolation and Culture

In this protocol, a procedure for aggregation of a non-small cell lung carcinoma (NSCLC) cell line, lung derived cancer-associated fibroblasts (CAF) as stromal cells, and blood-derived monocytes is described. This protocol may be adapted to other tumor cell lines, stromal, myeloid or other cell types.

General reagents: Trypsin-EDTA 0.05% (Gibco), Phosphate-buffered saline (PBS)

1. Tumor cell line: NCI-H157 (ATCC: #CRL-5802). Culture medium composition: RPMI media with 11 mM glucose and 2 mM GlutaMAX supplemented with 10% (v/v) FBS, 1% (v/v) penicillin/streptomycin, 12 mM HEPES, 1 mM sodium pyruvate, and 0.1 mM of non-essential amino acids (all from Life Technologies).

2. Lung derived cancer-associated fibroblasts (CAF) [11, 24–26] (Madar et al. 2009; Rudisch et al. 2015). Culture medium composition: RPMI media with 11 mM glucose and 2 mM GlutaMAX supplemented with 10% (v/v) FBS and 1% (v/v) penicillin/streptomycin (all from Life Technologies).

3. Blood-derived monocytes: peripheral blood mononuclear cells (PBMC) isolated from buffy coats. Reagents and solutions for the isolation procedure: phosphate buffered saline (PBS) containing 2% (v/v) FBS and 2 mM EDTA, ACK Lysis Buffer (Gibco), Lymphoprep™ and EasySep™ human monocyte isolation kit (both from Stemcell technologies). Culture medium composition: RPMI media with 11 mM glucose and 2 mM GlutaMAX supplemented with 10% (v/v) FBS and 1% (v/v) penicillin/ streptomycin (all from Life Technologies).

4. Coculture medium composition: RPMI media with 11 mM glucose and 2 mM GlutaMAX supplemented with 10% (v/v) FBS and 1% (v/v) penicillin/streptomycin (all from Life Technologies).

2.4 3D-3-Cell Culture

1. Stirred tank vessels: 125 mL stirred-tank spinner vessels with flat centered cap and angled side arms (Corning) or equivalent.
2. Magnetic stirring plate.

2.5 Culture Characterization

1. Fluorescein diacetate (Sigma).
2. TO-PRO® 3 (Invitrogen).
3. Fixation solution: Formaldehyde 4% (w/v) with 4% (w/v) sucrose in PBS.
4. Antibodies for characterization of immune cells by immunohistochemistry: CD45 (clone 2B11 þ PD7/26 from Agilent/DAKO), CD68 (clone PGM1 from Agilent/DAKO), and CD163 (clone 10D6 from Leica Biosystems).
5. Antibodies for detection of immune cell surface markers by flow cytometry: Anti-CD45 (BD Biosciences), Anti-CD206 (Biolegend), Anti-CD80 (Biolegend), and Anti-CD163 (R&D Systems).

3 Methods

All reagent preparation and cell culture handling should be carried out in a biological safety II cabinet under sterile conditions.

3.1 Tumor Cell Aggregation

Cell aggregation is a cell line dependent process. Culture parameters such as inoculum density, stirring rate, and time of aggregation must be adapted for aggregation of each different cell line. These parameters can also be adapted to attain different aggregation profiles, such as spheroids of different size and different spheroid concentration. This protocol describes aggregation of the NCI-H157 cell line.

1. Expand NCI-H157 as monolayers using standard culture protocols. Detach confluent cell monolayers (e.g., using trypsin or other enzymatic methods) and determine cell concentration and viability using the Trypan blue staining (or equivalent).
2. Prepare a single-cell suspension of 80 mL of complete medium at a cell concentration of 3×10^5 viable cell/mL and inoculate in a 125 mL stirred-tank spinner vessel. Maintain the spinner vessel at a stirring rate of 80 rotations per minute (rpm) for 24 h.
3. To monitor cell aggregation, collect 1 mL sample from the stirred-tank vessel culture under constant agitation in sterile conditions, transfer to a 24-well plate and observe in a brightfield microscope (*see* **Note 1**).
4. A progression from single cell to duplets and triplets is observed within the first hour of culture. After 24 hours,

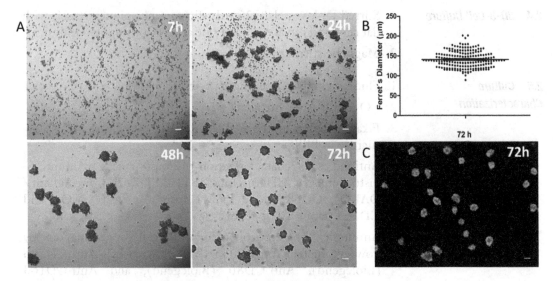

Fig. 2 NCI-H157 spheroid viability and aggregation dynamics during 3D culture. (**a**) Aggregation dynamics of tumor cells NCI-H157 over time (7, 24, 48, and 72 h). Scale bars represent 100 μm. Within the first 7 h, cell clusters are formed. At 24 h, the diameter of the cell aggregates increase, and at 72 h the morphology is loose. At 72 h, compact spheroids can be observed. (**b**) Average spheroid diameter after 72 h. (**c**) Live/dead assay of NCI-H157 spheroids 72 h after inoculation: FDA (green)—live cells; TO-PRO® 3 (red)—dead cells

irregular-shaped multicellular aggregates are formed (Fig. 2). At this point, add 40 mL of pre-warmed complete medium to the spinner vessel and increase the stirring rate to 100 rpm. Maintain the culture at these conditions until 72 h post-inoculation.

5. At 72 h post-inoculation, collect samples under constant agitation in sterile conditions for determination of cell viability, cell concentration, spheroid concentration, and spheroid diameter.

6. For determination of cell viability: transfer 0.5 mL of sample to a 24-wellplate and incubate spheroids for 5 min with 20 μg/mL fluorescein diacetate, which stains viable cells, and 1 μM of TO-PRO® 3, a DNA dye that stains cells with compromised membrane. Image spheroids using a fluorescence microscope equipped with the appropriate filters.

7. For determination of cell concentration and viability: collect 1 mL of sample to a 1.5 mL Eppendorf tube, centrifuge for $50 \times g$ for 1 min. Discard the supernatant, add 1 mL of PBS and repeat the centrifugation procedure described above. Discard the supernatant and add 0.3 mL of Trypsin. Incubate for 5 min at 37 °C. After incubation, gently pipette up and down the cell suspension and add 0.7 mL of complete medium. Determine cell concentration and viability using the Trypan blue staining (or equivalent).

8. For determination of the spheroid concentration: collect at least 1 mL of sample, divide in wells in a 96-well plate, visualize using a bright-field microscope and count the number of spheroids.

9. For determination of spheroid's diameter: use the open source software for image analysis ImageJ (or equivalent).

3.2 Peripheral Blood-Derived Monocytes Isolation

All procedures with blood should be performed at room temperature (RT).

1. Collect the peripheral blood and dilute it in two volumes of PBS containing 2% (v/v) FBS and 2 mM EDTA (ratio 1:2).

2. Add 15 mL of Lymphoprep™ to a 50 mL falcon tube and add 30 mL of the diluted blood on top of Lymphoprep™ (*see* **Note 2**).

3. Set the centrifuge brake off and centrifuge the 50 mL tubes at 950 × *g* for 25 min at RT.

4. After density gradient centrifugation, three layers will be distinctly identified (*see* **Note 3**). Collect the middle fraction containing the mononuclear cells (*see* Fig. 1b).

5. Transfer the middle fraction to new falcon tubes and add 30 mL of PBS containing 2% (v/v) FBS and 2 mM EDTA. Homogenize the cell suspension and centrifuge for 10 min at 400 × *g*. Repeat this step twice to remove platelets (*see* **Note 4**).

6. For erythrocyte removal, incubate the cell pellets with 5 mL of ACK Lysis Buffer for 5 min at 37 °C. Add 40 mL of PBS containing 2% (v/v) FBS and 2 mM EDTA and centrifuge for 10 min at 400 × *g* (*see* **Note 5**).

7. Collect all pellets in a single tube with 45 mL of PBS containing 2% (v/v) FBS and 2 mM EDTA and determine PBMC concentration and viability by Trypan Blue staining (or equivalent).

8. Isolate the monocytes from PBMC by negative selection through magnetic separation using EasySep™ human monocyte isolation kit, according to the manufacturer's instructions.

9. Determine the monocyte concentration and viability by Trypan blue staining (or equivalent).

3.3 Alginate Microencapsulation

1. Prior to the microencapsulation procedure, autoclave the following material: exchangeable nozzle, electrode, magnetic agitator, nozzle holder, tubes, aluminum vessel, and tweezers.

2. On the day of the procedure transfer the microencapsulation unit, syringe pump, and other material into the laminar flow hood and assemble the microencapsulation unit according to the manufacturer's instructions. Assemble the silicone tube to the nozzle holder.

3. Set the voltage to the desired unit (typically 4.7 Kw) (*see* **Note 6**).
4. Set the syringe pump to a flow rate of 10 mL/h.
5. Insert the magnet in the petri dish, dispense 50 mL of crosslinking solution, and turn the agitator for the magnet on.
6. In parallel, harvest the three cell types to be used in alginate microencapsulation (*see* **Note 7**).
7. To obtain CAF, expand monolayers using standard culture protocols. Detach confluent cell monolayers (e.g., using trypsin or other enzymatic methods) and determine cell concentration and viability using the Trypan blue staining (or equivalent).
8. Determine the cell suspension volume required to obtain 10×10^6 viable cells of CAFs.
9. After monocyte isolation, determine the cell suspension volume required to obtain 10×10^6 viable cells.
10. Transfer both cell suspensions to the same tube and centrifuge at $300 \times g$ for 5 min, RT. After centrifugation, discard the culture medium and resuspend the cell pellet in PBS. Repeat the centrifugation step.
11. Discard the supernatant and gently resuspend the cell pellet in 1 mL of alginate 1.1% (w/v) (*see* **Note 8**).
12. Determine the spheroid culture volume required to obtain 10×10^6 viable cells; transfer the spheroid suspension to a 50 mL tube and centrifuge at $50 \times g$ for 1 min, RT. Gently resuspend the cell pellet in PBS. Repeat the centrifugation step.
13. Discard the supernatant and add the alginate solution containing CAF and monocytes to the tube. Gently resuspend the spheroid pellet to obtain a homogenous suspension of tumor spheroids and single cells of CAF and monocytes.
14. Transfer the solution to the syringe, remove air bubbles, and connect the syringe to the silicon tube.
15. Close the microencapsulation unit cabinet, turn the syringe pump and the voltage switch on (*see* **Note 9**).
16. After the alginate suspension is dispensed, maintain the alginate beads in the crosslinking solution for 7 min.
17. Recover the alginate beads and transfer to a 50 mL tube. Let the beads sediment by gravity, discard the crosslinking solution, and wash 3× with the washing solution (0.9% NaCl).
18. After the washing steps, resuspend the alginate beads in 20 mL of coculture medium.
19. Add 80 mL of pre-warmed coculture culture medium to a 125 mL stirred-tank spinner vessel.
20. Transfer the alginate beads to the stirred-tank spinner vessel and maintain the culture at 80 rpm.

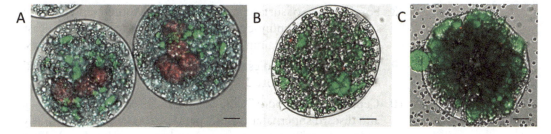

Fig. 3 Characterization of the microencapsulated 3D-3-coculture. (**a**) Alginate microcapsules of the cocultures after microencapsulation, visualized by fluorescence microscopy. The cellular types are labeled by tdTomato (red)—NSCLC spheroids; GFP (green)—CAF; Cell tracker™ (blue)—myeloid cells. (**b, c**) Live/dead assay of 3D-3-cocultures in the first (**b**) and third (**c**) week of culture: FDA (green)—live cells; TO-PRO® 3 (red)—dead cells. Cell viability is maintained over time. Scale bars represent 100 μm

3.4 3D-3-Culture Characterization

1. To monitor the microencapsulated 3D-3-coculture, collect 1 mL of sample from the stirred-tank vessel culture under constant agitation, transfer to a 24-well plate and observe in a bright-field microscope (Fig. 3).

2. To monitor cell viability, proceed as described above (Subheading 3.1, **step 6**).

3. To prepare samples for immunohistochemistry, discard the culture medium and wash the beads twice with PBS, by allowing sedimentation of the alginate beads. Incubate the alginate beads in the fixation solution for 20 min at RT. After incubation, wash the alginate beads twice with PBS. Prepare the samples for paraffin embedding and proceed with immunohistochemistry using macrophage specific markers (e.g., CD68, CD163).

4. To recover the cells for further characterization, collect the samples from the stirred-tank vessel to a tube, discard the culture medium and wash with PBS. After sedimentation of the alginate beads, discard the PBS and add chelating solution (10 mM HEPES pH 7.4 with 100 mM EDTA) to dissolve the alginate beads. Incubate for 5 min at RT (*see* **Note 10**).

5. To collect the non-infiltrated myeloid fraction, centrifuge the cell suspension for 2 min at 50 × g and transfer the supernatant to a new tube. Add the same volume of PBS and centrifuge for 10 min at 400 × g.

6. To collect infiltrated myeloid cells, incubate the pellet obtained in step 5 with Trypsin-EDTA for 5 min at 37 °C. Inactivate Trypsin-EDTA with complete medium.

7. Centrifuge tube for 10 min at 400 × g and wash once with PBS. Determine cell concentration and viability by trypan blue staining (or equivalent).

8. For detection of surface markers by flow cytometry, prepare the samples for immunocytochemistry by distributing at least 2×10^5 cells per staining in Eppendorf tubes or 96-well plates.
9. Prepare the antibodies mixtures according to manufacturers' instructions (*see* **Note 11**).
10. Centrifuge Eppendorf tubes or plates for 10 min at $400 \times g$ and discard supernatant. Resuspend pellets with the previously prepared antibody mixtures. Incubate samples for 1 h at RT. After incubation, centrifuge at $400 \times g$ for 10 min and wash 2× in PBS. Discard the supernatant and resuspend the pellets in the appropriate volume of PBS with 2% FBS. Samples are ready to be analyzed by flow cytometry.

4 Notes

1. It is critical to monitor the culture during the aggregation period. Aggregation in stirred systems is based on random collisions between single cells and is not a linear process. At some point, cell aggregates can start to fuse forming bigger spheroids. This process must be controlled so as not to reach the critical size where the diffusion of nutrient and O_2 to the inner part of the spheroids can no longer support cell viability and necrotic centers are formed. Monitoring spheroid size and concentration over time can help prevent these effects. This should be done by sampling the culture over time.
2. It is critical to dispense the blood fraction slowly to prevent mixing with Lymphoprep. At the end of this step two distinct fractions should be observed. If the two fractions are mixed, the separation between the layers obtained after density gradient centrifugation will be less clear and collection of the mononuclear cells will be more difficult.
3. The bottom layer contains erythrocytes and granulocytes; the middle layer contains mononuclear cells and top layer contains plasma (including platelets). When collecting the middle fraction, be careful not to disturb the top and bottom layers to prevent contamination with red blood cells or platelets.
4. Throughout the washes, the cell suspension will become translucid, indicating that platelet contamination is decreasing.
5. At this point, the pellet may still contain erythrocytes which were carried out at the step of middle layer removal and appear red. The incubation with ACK lysis solution can be repeated up to two times for improving erythrocyte removal. Monitor the color of the pellet to assess if the incubation step should be repeated.

6. The size of alginate beads is controlled by several parameters, such as the nozzle diameter, flow rate, viscosity of the alginate solution, and electrostatic voltage. These parameters can be adapted to the desired size of alginate beads. However, it must be considered that increasing the voltage may have a negative impact on cell viability.

7. To establish a homogeneous coculture, the alginate microencapsulation conditions were determined to obtain 1–2 spheroids per alginate microcapsule and the average diameter of the alginate microcapsules was approximately 700 μm. The ratio of tumor cells to stromal and myeloid cells in the 3D-3-coculture was 1:1:1 and 10×10^6 cells of each type were used. The cell ratios and quantities may be adapted.

8. The alginate solution is viscous; avoid bubble formation during resuspension of the cell pellet, since this can interfere with the electrostatically-driven microencapsulation procedure.

9. Monitor the microencapsulation procedure closely while the spheroid and single-cell mix in alginate is being dispensed. Ensure the newly formed alginate beads are under constant agitation and that the nozzle is not clogged. Prior to encapsulation, use an empty sterile syringe to insert air through the tubes and the nozzle to ensure that the system is not clogged, which will have detrimental effects on the size homogeneity of the capsules.

10. The volume of chelating solution must be adjusted to cover the volume of alginate beads. During alginate dissolution, monitor the cell suspension visually to confirm the alginate is completely dissolved and mix the cell suspension, if needed, to improve the dissolution. Incubation with the chelating solution for 5 min should not impact cell viability. However, longer incubations may be detrimental to the cells.

11. To ensure optimal staining, antibody titration may be carried out before to find the optimal antibody concentration.

References

1. Hay M, Thomas DW, Craighead JL et al (2014) Clinical development success rates for investigational drugs. Nat Biotechnol 32:40–51. https://doi.org/10.1038/nbt.2786
2. Ruggeri BA, Camp F, Miknyoczki S (2014) Animal models of disease: pre-clinical animal models of cancer and their applications and utility in drug discovery. Biochem Pharmacol 87:150–161. https://doi.org/10.1016/j.bcp.2013.06.020
3. Hickman J, Graeser R, de Hoogt R et al (2014) Three-dimensional models of cancer for pharmacology and cancer cell biology: capturing tumor complexity in vitro/ex vivo. Biotechnol J 9:1115–1128. https://doi.org/10.1002/biot.201300492
4. Santo VE, Rebelo SP, Estrada MF et al (2017) Drug screening in 3D in vitro tumor models: overcoming current pitfalls of efficacy readouts. Biotechnol J 12:1600505. https://doi.org/10.1002/biot.201600505
5. Weiswald LB, Bellet D, Dangles-Marie V (2015) Spherical cancer models in tumor biology. Neoplasia 17:1–15. https://doi.org/10.1016/j.neo.2014.12.004

6. Hirt C, Papadimitropoulos A, Mele V et al (2014) "In vitro" 3D models of tumor-immune system interaction. Adv Drug Deliv Rev 79:145–154. https://doi.org/10.1016/j.addr.2014.05.003
7. Nyga A, Neves J, Stamati K et al (2016) The next level of 3D tumour models: immunocompetence. Drug Discov Today 21:1421–1428. https://doi.org/10.1016/j.drudis.2016.04.010
8. Alemany M, Semino CE (2014) Bioengineering 3D environments for cancer models. Adv Drug Deliv Rev 79–80:40–49. https://doi.org/10.1016/j.addr.2014.06.004
9. Santo VE, Estrada MF, Rebelo SP et al (2016) Adaptable stirred-tank culture strategies for large scale production of multicellular spheroid-based tumor cell models. J Biotechnol 221:118–129. https://doi.org/10.1016/j.jbiotec.2016.01.031
10. Asghar W, El Assal R, Shafiee H et al (2015) Engineering cancer microenvironments for in vitro 3-D tumor models. Mater Today 18:539–553. https://doi.org/10.1016/J.MATTOD.2015.05.002
11. Stock K, Estrada MF, Vidic S et al (2016) Capturing tumor complexity in vitro: comparative analysis of 2D and 3D tumor models for drug discovery. Sci Rep 6:28951. https://doi.org/10.1038/srep28951
12. Estrada MF, Rebelo SP, Davies EJ et al (2015) Modelling the tumour microenvironment in long-term microencapsulated 3D co-cultures recapitulates phenotypic features of disease progression. Biomaterials 78:50–61. https://doi.org/10.1016/j.biomaterials.2015.11.030
13. Quail DF, Joyce JA (2013) Microenvironmental regulation of tumor progression and metastasis. Nat Med 19:1423–1437. https://doi.org/10.1038/nm.3394
14. Gajewski TF, Schreiber H, Fu Y-X (2013) Innate and adaptive immune cells in the tumor microenvironment. Nat Immunol 14:1014–1022. https://doi.org/10.1038/ni.2703
15. Sica A, Allavena P, Mantovani A (2008) Cancer related inflammation: the macrophage connection. Cancer Lett 267:204–215. https://doi.org/10.1016/j.canlet.2008.03.028
16. Solinas G, Germano G, Mantovani A, Allavena P (2009) Tumor-associated macrophages (TAM) as major players of the cancer-related inflammation. J Leukoc Biol 86:1065–1073. https://doi.org/10.1189/jlb.0609385
17. Xue J, Schmidt SV, Sander J et al (2014) Transcriptome-based network analysis reveals a spectrum model of human macrophage activation. Immunity 40:274–288. https://doi.org/10.1016/j.immuni.2014.01.006
18. Aras S, Raza Zaidi M (2017) TAMeless traitors: macrophages in cancer progression and metastasis. Br J Cancer 117:1583–1591. https://doi.org/10.1038/bjc.2017.356
19. Lavin Y, Kobayashi S, Leader A et al (2017) Innate immune landscape in early lung adenocarcinoma by paired single-cell analyses. Cell 169:750–765.e17. https://doi.org/10.1016/j.cell.2017.04.014
20. Sun Y (2015) Translational horizons in the tumor microenvironment: harnessing breakthroughs and targeting cures. Med Res Rev 35:408–436. https://doi.org/10.1002/med.21338
21. Mok S, Koya RC, Tsui C et al (2014) Inhibition of CSF-1 receptor improves the antitumor efficacy of adoptive cell transfer immunotherapy. Cancer Res 74:153–161. https://doi.org/10.1158/0008-5472.CAN-13-1816
22. Gordon SR, Maute RL, Dulken BW et al (2017) PD-1 expression by tumour-associated macrophages inhibits phagocytosis and tumour immunity. Nature 545:495–499. https://doi.org/10.1038/nature22396
23. Rebelo SP, Pinto C, Martins TR et al (2018) 3D-3-culture: a tool to unveil macrophage plasticity in the tumour microenvironment. Biomaterials 163:185–197. https://doi.org/10.1016/j.biomaterials.2018.02.030
24. Haubeiss S, Schmid JO, Mürdter TE et al (2010) Dasatinib reverses cancer-associated fibroblasts (CAFs) from primary lung carcinomas to a phenotype comparable to that of normal fibroblasts. Mol Cancer 9:168. https://doi.org/10.1186/1476-4598-9-168
25. Madar S, Brosh R, Buganim Y, Ezra O, Goldstein I, Solomon H, Kogan I, Goldfinger N, Klocker H, Rotter V (2009) Modulated expression of WFDC1 during carcinogenesis and cellular senescence. Carcinogenesis 30(1):20–27. https://doi.org/10.1093/carcin/bgn232
26. Rudisch A, Dewhurst MR, Horga LG, Kramer N, Harrer N, Dong M, van der Kuip H, Wernitznig A, Bernthaler A, Dolznig H, Sommergruber W (2015) High EMT signature score of invasive non-small cell lung cancer (NSCLC) cells correlates with NFκB driven colony-stimulating factor 2 (CSF2/GM-CSF) secretion by neighboring stromal fibroblasts. PLoS One 10(4):e0124283. https://doi.org/10.1371/journal.pone.0124283

Chapter 9

Considerations in Developing Reporter Gene Bioassays for Biologics

Jamison Grailer, Richard A. Moravec, Zhijie Jey Cheng, Manuela Grassi, Vanessa Ott, Frank Fan, and Mei Cong

Abstract

The establishment of a robust and reproducible functional bioassay that reliably measures drug potency while ascertaining its mode of action is essential in biologic drug development. Here we describe a simple bioluminescent reporter gene bioassay for assessing biologics targeting immune checkpoints without the complexity and variability of more traditional assay systems. This chapter provides an overview of key considerations in reporter gene bioassay design and optimization, as well as development of thaw-and-use cells as an assay reagent for biologic QC lot release.

Key words Biologics, Immunotherapy, Reporter bioassay, Thaw-and-use cells, Quality control, Drug potency

1 Introduction

Vaccines, blood and blood components, allergenics, tissues, and recombinant therapeutic proteins are biological products, comprising more than 50% of new drugs in development. Among the different classes of biologics, monoclonal antibodies (mAbs) represent the most broadly developed therapeutics. The continued increasing R&D investment in biologic drug development reflects their specificity and thus fewer off-target effects, as well as longer exposure compared to small-molecule-based therapeutics.

Biologics have emerged as promising new approaches to immunotherapy, which aim to utilize the patient's own immune system to combat diseases such as cancer and autoimmunity. Some of these approaches include recombinant cytokines, immune checkpoint

---

The original version of this chapter was revised. The correction to this chapter is available at https://doi.org/10.1007/978-1-0716-0171-6_14

Jamison Grailer, Richard A. Moravec, and Zhijie Jey Cheng contributed equally to this work.

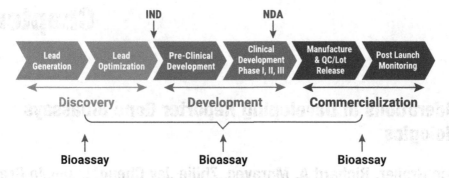

Fig. 1 Bioassays can fit into every stage of biological drug development

mAbs, bispecific molecules, antibody-drug conjugates, and chimeric antigen receptor (CAR) T cell therapies. However, the inherent complexity, functional heterogeneity, and sensitivity of biologics to external storage conditions present unique challenges in drug discovery and development, and the development and manufacture of these products must be tightly controlled to ensure consistent purity, potency, efficacy, and safety. To address these challenges, functional, mechanism of action (MOA)-based bioassays are used throughout the biologics drug discovery and development workflow to screen, characterize drug MOA, and monitor product bioactivity, potency, and stability (Fig. 1).

A bioassay uses living material (animal, plant, tissue, or cells) to measure the biological activity of a substance. In biopharmaceutical drug development, bioassays are typically cell-based assays that are used to measure the bioactivity and potency of a biologic drug. In biologic drug development, it is challenging but critical to establish a functional bioassay that meets the essential quality attributes for measuring drug potency as described by regulatory guidelines.

Potency, a measurement of the strength of biological activity, is a functional measure of the tertiary/quaternary structure of a biologic drug as it relates to its therapeutic MOA. It is assessed in a bioassay by comparing the dose–response curve of the test material with that of a reference standard in a multiwell plate-based assay format [1]. This property is a critical parameter of drug product quality release testing and is also used to monitor drug stability, demonstrate product comparability after a manufacturing process change, and to assess lot-to-lot consistency during normal manufacturing operations. It is, therefore, critical that potency bioassays developed for use in biologics manufacturing and QC lot release reflect the MOA of the drug.

Traditional approaches to developing potency bioassays have relied on animal models and primary cells. These model systems provide biological relevancy for characterizing a drug's MOA, but they are challenging to implement in a quality-controlled manufacturing environment due to variability in the sourcing of primary cells, complex assay protocols, and limited availability of qualified reagents.

In recent years, bioluminescent reporter gene bioassays have been developed and validated for use in biologic drug manufacturing and QC lot release. Reporter gene bioassays can be designed to reproduce a biologic drug MOA while providing a measure of drug potency without the assay complexity and variability of more traditional model systems. This chapter provides an overview of key considerations in reporter gene bioassay design, clone selection, assay optimization, and qualification for manufacturing and QC lot release.

2 Assay Design

2.1 Choosing a Cell Background

The first consideration in designing a reporter gene bioassay is to choose an appropriate cell background. The cells must express biologically relevant signaling molecules and pathways to recapitulate the in vivo functional response targeted by the drug. Primary cells typically meet these criteria but are challenging to implement in higher-throughput and quality-controlled environments due to their variability and complex assay protocols. Many biologics drugs are designed to recognize therapeutic targets expressed on the cell surface, and, therefore, a cell line endogenously expressing the target receptor is a good option. For example, a HER2$^+$ breast cancer cell line is a good option for measuring the potency of an anti-HER2 biologic drug. However, many immortalized cell lines endogenously express multiple receptors that can potentially activate the same signaling pathway. Therefore, it is important to demonstrate assay specificity by showing that the reporter gene response is dependent upon engagement of the target receptor.

Transformed cell lines are a less variable alternative to primary cells that are easier to develop into a quantitative and reproducible bioassay. If a cell line endogenously expressing a receptor target of interest is not available, or if the cell line shows a significant nonspecific response, a biologically less relevant cell line genetically engineered to express a specific target receptor can be used. Regardless of which approach is taken, a bridging study is typically required to demonstrate that the reporter gene bioassay exhibits the engagement of the target upon the binding of the biologics and an equivalent response compared to a primary cell-based assay.

2.2 Selecting a Genetic Reporter

The Luciferase Assay System is an extremely sensitive and rapid reagent for the quantitation of firefly luciferase. Linear results are seen over at least eight orders of magnitude of enzyme concentration, and less than 10^{-20} moles of luciferase can be measured under optimal conditions. Therefore, luciferase reporters are widely used in cell-based assays due to their large dynamic range, homogeneity, and simple add-and-read assay format.

In addition, a biologically relevant promoter that exhibits rapid and robust activation in response to the biologic product with minimal nonspecific activation should be identified. A biologic product may be able to activate multiple intracellular signaling pathways, resulting in the activation of several transcription factors and promoters. In this situation, a promoter that is most directly coupled to the target pathway is preferred. For example, if a cytokine drug is designed to promote cell proliferation, a promoter that directly contributes to cell growth would be the most appropriate choice. Criteria for evaluating candidate promoters include activation kinetics, response fold induction, and relevant EC_{50} values.

2.3 Identifying a Positive Control

In order to make comparisons within and between bioassay optimization runs, a positive control biologic must be identified and used consistently throughout bioassay development. A biologic or drug product could be used as the positive control.

The positive control biologic should be stable at the recommended storage conditions (typically stored in aliquots at $-80\ °C$), and multiple lots should be tested to ensure lot-to-lot consistency. A positive control is not the same as a reference biologic, which is manufactured according to the same processes as a biologic product and used to determine the relative potency of a biologic in later-stage development and manufacturing QC lot release. However, if a reference biologic is available, it can be used as a positive control for bioassay development and during the later stages of bioassay validation, system suitability testing, and potency determination for lot release.

3 Assay Feasibility Studies

The goal of feasibility studies is to demonstrate proof-of-concept that the bioassay will perform as expected using the cell background and genetic reporter identified during the assay design phase. Thorough characterization and demonstration of the underlying biology of a bioassay is essential to avoid unnecessary time and cost spent in the later phases of bioassay optimization and qualification.

Feasibility studies using genetically engineered cell lines can be performed by transient transfection or through the creation of a stable cell pool. Transient transfection is a good option if the cells can be easily transfected with high efficiency and the target receptor is expressed at a relatively high level. Factors such as cell background, expression construct, transfection method, and underlying biology may all impact whether feasibility studies can be accomplished using transiently transfected cells. In some instances, a stable cell pool will need to be established using antibiotic selection. When using a new cell line to create a stable cell pool, it is important to generate an antibiotic kill curve to determine the optimal concentration of antibiotic to use for selection. A concentration of antibiotic strong enough to kill the nontransfected cells but not too strong to kill all the cells should be used. Antibiotic selection typically requires 2–5 weeks to complete depending on the cell background and antibiotic selection marker.

Initial functional studies should be performed to generate a 10-point dose–response curve of a positive control biologic. At this early stage of proof-of-concept study, a perfect curve with 2–3 points at upper asymptote and 2–3 points at lower asymptote is not required. However, a lack of dose-dependent response may be the result of low receptor expression or a nonfunctional genetic reporter. The generation of a stable cell pool does not necessarily equate to enough receptor expression for bioassay development, and, therefore, receptor expression should be measured directly by flow cytometry or other methods. If receptor expression is determined to be low, the cells can be further sorted to obtain a population of higher-expressing cells. Importantly, while high target-receptor expression may be desirable for some bioassay designs (e.g., to measure the activity of a soluble ligand or antibody), other bioassays that require complex interactions between multiple cell types and receptor–ligand pairs may benefit from lower receptor expression. In these cases, it is recommended to sort multiple populations of cells with varying receptor expression levels for functional testing. Finally, if no dose–response curve is observed and relatively high target–receptor expression is demonstrated, it is possible that the genetic reporter is nonfunctional. To assess this possibility, soluble compounds that nonspecifically activate promoter elements can be tested.

As noted above, some bioassay designs that use multiple cell types and receptor–ligand pairs may benefit from lower receptor-level expression. High target-receptor expression can lead to increased basal activity and a reduced assay window due to the dynamic equilibrium between the active and inactive forms of the receptor. Therefore, while target–receptor expression can be informative to interpret feasibility data, the functional assay response must be the primary criterion for decision making.

In summary, the goal of feasibility studies is to demonstrate that the bioassay cell line can yield a dose-dependent response using a positive control biologic. Even a less than twofold positive response is typically sufficient to move toward further development of the bioassay. If a dose–response curve cannot be generated using a positive control biologic, even when the target–receptor expression and reporter function are confirmed, alternative assay design strategies (e.g., alternative cell line or promoter) should be considered.

4 Cell Line Generation and Clone Stability Testing

4.1 Selecting Cell Clones

The cell line developed for use as a functional bioassay should consist of a single cell clone. This ensures stable integration of genetic elements, reduces genetic drift or loss of the engineered content, and results in more reproducible assay performance over time.

Single cell clones are typically generated by limiting dilution where a suspension of cells is diluted and dispensed into 96-well (or higher-throughput) plates such that each well contains an average of at most one cell per well. The cells are then cultured and expanded resulting in a population of cells derived from a single cell clone. Limiting dilution is relatively easy to perform but does result in wells with either no cells or more than one cell. Wells with more than one cell may not be genetically identical and may result in the generation of an unstable population. Multiclonal wells of adherent cells can easily be identified by visual inspection, but suspended cells remain a challenge.

If ectopic expression of the target receptor is used, another approach used to generate single cell clones involves labeling the target of interest with fluorescently labeled antibody and using fluorescence-activated cell sorting (FACS) to sort the labeled cell population into a 96-well plate at a density of one cell per well. This technique requires a cell-labeling step and subjects the cells to high pressure during sorting, which can be overly stressful for some cell types when culturing from a single cell. Thus, not all wells containing single cells expand into clonal populations. Success rates will vary depending on the growth characteristics of the cell line and overall health of the cells prior to sorting. If antibody labeling is performed as part of the FACS protocol, it is good practice to first sort a pool of cells based on the antibody label, let the cells recover in culture, and then blind sort (e.g., without antibody labeling) single cell clones into plates. This approach limits cell handling on the day of single-cell cloning and increases the success rate of clonal cell expansion.

After limited dilution cloning (or FACS to sort single cells), the cells are cultured in multiwell plates (typically 96 wells) until the cell population has expanded to at least 20–30% confluent to allow initial functional screening. During the initial phase of cell culture and expansion, it is typical to see a wide range of cell growth rates between individual clones. Some clones may grow as well as their parental cell line while others may stop growing.

An initial screen can be performed by replicating individual wells into parallel plates and measuring the functional response with or without one single concentration of a positive control biologic. Functional responses can be categorized according to whether they show a low, medium, or high response.

Luminescence observed during the initial screen can vary dramatically among different cell clones. Clones showing high basal relative luminescent unit (RLU) signal and producing a high-fold induction in response to a positive control biologic are the best candidates for further bioassay development. Clones showing higher basal RLU will result in better assay performance as measured by percent coefficient of variations (% CV). In general, basal RLU should be at least 50- to 100-fold higher than instrument background (wells with medium only). However, if low basal RLU is generally observed in the initial screen, clones with higher-fold induction will be chosen for further assay optimization to increase the basal RLU.

When clones are further expanded, full dose–response curves can be tested on a subset of clones. When comparing clone functional responses, cells should be seeded at the same cell density so that their yields are comparable on the day of harvest for the assay. Cells should be plated at the same number of cells per well so that comparisons can be made. Full-dose titration of the ligand or biological product will give insight into the EC_{50}, maximum fold induction, shape of the response curve (hillslope and curve fit parameters), and overall luminescence intensity. Even at this early stage of bioassay development, some assays respond strongly using normal serum concentrations whereas others benefit from low percentage or alternative sera (e.g., charcoal-stripped serum).

Multiple criteria should be taken into consideration when selecting a cell clone for bioassay development. For reporter gene bioassays, luminescence fold-induction in response to a positive control biologic is a key attribute. However, other parameters such as signal background, peak signal intensity, and induction time should also be considered.

If a bioassay is developed for use with a blocking antibody to neutralize an agonist, the agonist concentration used for potency measurement of the blocking antibody and slope response of the agonist dose–response curve will impact the EC_{50} of the test blocking antibody. Higher concentrations of agonist will result in a higher EC_{50} of the blocking antibody. It is important to mimic a historically acceptable EC_{50} range when comparing clone responses.

If ectopic receptors need to be added to the parent cell, flow cytometry can be used to determine relative expression levels of different clones, correlate response to expression level, and even identify clones that most likely are not truly clonal. As shown in Fig. 2, double peaks and broad peaks with a "shoulder" of fluorescence intensity often indicate a mixed culture of clones. While this "clone" may temporarily meet the bioassay needs, it may ultimately fail when undergoing passage stability.

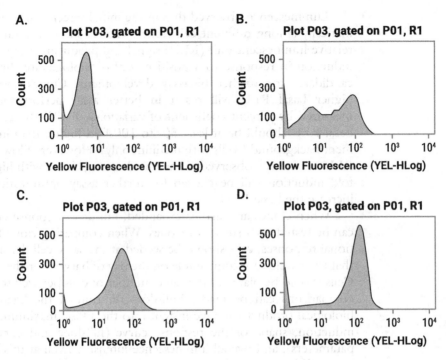

Fig. 2 Histograms of three different clones following FACS staining. (**a**) Isotype control. (**b**) Double peak indicating mixed culture. (**c**, **d**) Two clones demonstrating slightly different levels of expression

4.2 Performing Clone Stability Testing

"Cell passage" refers to the number of cell population doublings, which accounts for variable growth rates and is not affected by the number of times the cells are passaged in a week. It is important to establish the length of passage stability and acceptable level of loss or change, which will depend on the individual requirements of a clone. Long-term clonal stability ensures consistent cellular functional responses throughout passages in terms of EC_{50} response, fold induction, overall luminescence, cell growth, and receptor expression during a defined amount of time in culture. Functional instability can manifest as a decrease in luminescence while maintaining fold induction or as a decrease in fold induction (Fig. 3). At a minimum, cell clones must be stable enough to sequentially prepare seed stocks, master cell bank, and working cell bank.

Stability studies can begin after initial functional screening and selection of a limited number of clones that exhibit assay specificity and good assay response. Cells are maintained under full antibiotic selection pressure as indicated during preliminary kill curve tests with the parental cell line. Cell culture medium with freshly supplemented antibiotics from reliable suppliers will help individual clones maintain their original characteristics and minimize genetic drift. Sufficient banks should be made for each testing clones at passages as early as possible to serve as a source for subsequent seed stock preparation or "backup" cells.

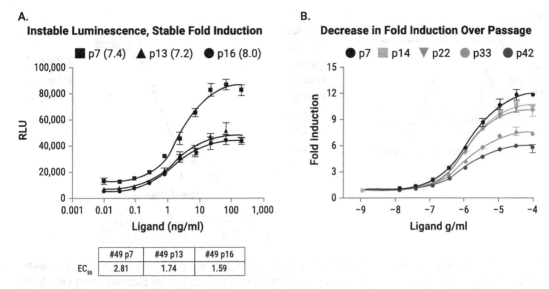

Fig. 3 Examples of clonal instability. (**a**) Decrease in luminescence across passages 7–16 (stable fold induction noted). (**b**) Steady fold induction decrease across passages 7–42

Several strategies can be used to prepare a candidate cell clone for stability testing. Cells for each clone representing one or more cell passages can be maintained in parallel over time as active cell cultures. Cells will be harvested and frozen every 5–6 passages to generate a series of staggered passages. Cells frozen at different passages will then be thawed and grown in parallel for at least 2 weeks, followed by side-by-side functional measurement of basal luminescence (measure of reporter gene stability), signal fold induction (measure of receptor expression), and EC_{50}. Comparing functional response data from different days is not a preferred approach as many variables in assay and cell culture conditions will contribute to the assay results and complicate data interpretation. To confirm functional results, additional tests such as flow cytometry analysis for receptor expression are recommended using the same staggered passage cultures. Flow cytometry and functional response can be used in tandem to demonstrate a stable clone (Fig. 4). Other cell culture characteristics, such as cell population doubling time and cell morphology, can also be noted during this expansion duration and contributed to final clone selection.

5 Assay Optimization

Optimization of bioassays is important to ensure the best possible sensitivity, signal-to-noise ratio, and assay window. The bioassay must also be reproducible, ensuring that small day-to-day variations do not significantly impact the results. Several critical factors during bioassay optimization are considered below.

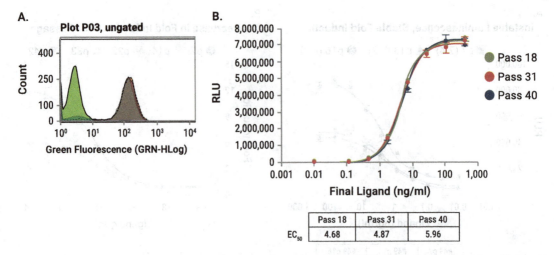

Fig. 4 Stable functional response confirmed with surface expression. (**a**) Cells from passages 18 to 40 were analyzed for receptor expression by flow cytometry; isotype control included. (**b**) Bioassay using cells from passages 18 to 40

5.1 Standardizing Assay Reagents

Quality cell culture and assay reagents are critical for consistent bioassay performance. It is prudent to identify reliable suppliers of quality sera, media, and cell culture supplements and establish assay media formulation during bioassay development. Sera, in particular, can have significant impact on cell growth and assay performance depending on grade, region of origin, and supplier. Always anticipate supply constraints and test multiple media and sera suppliers to ensure consistent performance of the bioassay.

5.2 Culturing Cells

Passaging of cells according to a defined schedule is integral to consistent assay performance. For most bioassay cell lines, a Monday–Wednesday–Friday–Monday schedule is recommended. Optimization of culture conditions often results in consistent growth rate (doubling time) with high cell viability at the time of harvest.

When expanding cells, seed them into flasks and measure the seeding density. Seed suspension cells at a defined cell density (number of cells per mL) in a set volume of medium per flask size. For adherent cells, use cell numbers per cm^2 for calculating cell seeding density in each flask. Volume differences between flasks can affect gas exchange and cell performance. Therefore, a standardized media volume per flask (e.g., 20 mL per T75 flask) is recommended. Cell seeding and harvesting densities are best tested empirically for effects on cell performance. Seeding cells too densely results in overgrown cultures that typically are detrimental to assay performance. On the other hand, seeding cells too sparsely can result in increased costs. Test a range of seeding densities during a standardized 2- or 3-day passage to determine the best range so that variations in cell seeding density (e.g., inaccurate cell count at seeding) have as little impact on reproducibility of the bioassay as feasible. Record cell harvesting density, doubling rate, and viability at the time of harvest as integral data for the bioassay.

5.3 Plating Cells

While optimizing bioassay conditions, it is important to test a range of cell numbers per well in order to determine how signals are detected. Since instrument detection sensitivity can vary, it is important to consider the signal-to-background ratio to determine the level of signal that can be obtained using a positive control above the background noise of the machine. Background noise is indicative of the lowest sensitivity that can be determined for the bioassay. If the bioassay includes a coculture of multiple cell types (e.g., effector and target cells), cell–cell interaction and cell ratio can have a significant impact on the assay window and sensitivity. Test a range of cell ratios to ensure optimal performance of a bioassay with cell cocultures. For adherent-suspension cell interactions, adherent cells can be plated overnight to reach 80–100% confluency the next day and interact with 50,000 suspension cells per well in a 96-well plate. For suspension–suspension cell interactions, a 1:1 ratio (e.g., 50,000 cells for each cell type) in each well of a 96-well plate will be a good starting point.

5.4 Induction Time

Assay incubation time can also have a significant impact on reporter-based bioassay performance. Induction time should be optimized by performing a time course experiment at 4, 5, and 6 h, or overnight (18–24 h) to determine the maximal performance and assay robustness. When harvesting and staging the bioassay with adherent cells, significant response differences may be observed when using traditional trypsin or weaker enzymatic alternatives as surface proteins and receptors can be temporarily damaged. Certain bioassay cells may need an overnight "recovery" after plating to achieve an optimal and consistent response. A solid understanding of the nuances of a cell clone will greatly improve the odds of a successful bioassay or its transfer to another facility.

5.5 Assay Buffer and Media

For cell culture media, a common practice is to use a general, well-established medium composition according to the cell origin document and recommendations from commercial cell line providers (ATCC, DSMZ) and supplement it with fetal bovine serum (FBS) (e.g., normal, heat-inactivated, or gamma-irradiated) and together with nutrients that support cell growth. Since FBS is a natural product without clearly defined composition, test sera from different commercial sources, and even different lots from the same commercial source, are used to confirm that they do not cause significant assay variation. The optimal percentage of FBS is typically between 0.5% and 10%. In some situations, components of FBS can interfere with the bioassay, in which case the use of dialyzed FBS should be considered. Being able to lower FBS

5.6 Plate Edge Effects

concentration in assay buffer without impacting cell health, luminescence signal, and assay robustness will help to minimize variations introduced by different FBS lots.

Plate edge effect is an inconvenient phenomenon caused by more media evaporation along the edges of the microplate during incubation. The edge effect can cause many problems, including varying volumes and concentrations, which can alter cell viability and assays results. In general, only the inner 60 wells of a 96-well plate are used for most bioassay applications to avoid plate edge effects on data. If a bioassay has a short induction time (e.g., less than 5–6 h) plate edge effects are of less concern. To test for edge effects, generate a plate "heat map" where each well gets the same treatment (typically the EC_{50} of the test biologic). Luminescence across the plate will indicate whether there are significant position-dependent effects.

5.7 Hook Effect

Occasionally, upon careful inspection of the fitted data, a "hook effect" may be noticed at the highest concentration(s) tested that may impact the curve fit. To troubleshoot, confirm the stock biologic does not contain a toxic component (such as sodium azide, detergent, or stabilizer) which may carry over to the first dilution (highest concentration) sample tested, thereby impacting the cell health and decreasing luminescence. Sometimes antibodies are prone to hook effects when used at high concentrations. Preparing a series of tightly spaced data points across a narrow concentration range can shed insight as to where along the concentration range the hook appears. Altering the starting dilution concentration, with anticipation of the highest potency desired, will typically resolve the problem.

5.8 Design of Experiments (DoE)

Assay optimization can be performed by assessing one parameter at a time. However, interplay between assay parameters can significantly impact the robustness and repeatability of a bioassay. Therefore, multifactorial analysis performed by varying multiple experimental parameters in a single experiment is recommended for assay optimization. Design of experiment (DoE) is a commonly applied method of multifactorial analysis that is used to define the relationships between assay parameters that impact assay output. It is extensively used for the implementation of Quality by Design in both research and industrial settings. DoE can be simple or complex, depending on the application and number of critical assay parameters being evaluated. Figure 5 shows an example of a simple DoE experiment used to demonstrate the robustness of a bioassay previously optimized one parameter at a time.

In this example, five assay parameters (two conditions per parameter, one being the final optimized condition) were tested in a single experiment (Fig. 5a). Statistical analysis software (JMP)

Developing Reporter Gene Bioassays for Biologics 143

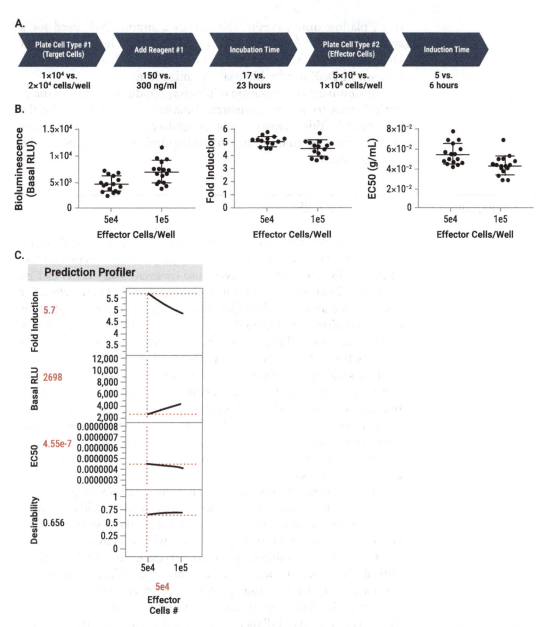

Fig. 5 Example DoE design and data analysis. (**a**) Five critical assay parameters were tested in a single DoE experiment. (**b**) Aggregate data of differences in relative luminescence (RLU), assay window (fold induction), and EC$_{50}$ of a control biologic when the number of effector cells is varied. Bars show mean ± standard deviation. (**c**) Example of data analysis using JMP software. Shown are the effects of effector cell number on fold induction, RLU, and EC$_{50}$. The desirability index is shown where fold induction and RLU are equally weighted

was used for the assay setup design and data analysis (Fig. 5c). The assay was successful as it demonstrated a positive response to the biologic in all cases (Fig. 5b), but some assay parameters, such as cell number, had bigger effects on the assay readout than others (e.g., plating time). Again, this is only a simple DoE used for a specific purpose. A more detailed explanation of how to perform DoE is outside of the scope of this chapter. Consult a biostatistician to assist in DoE use for bioassay optimization.

In summary, optimization of bioassay conditions includes many critical steps to achieve consistent bioassay performance. The list presented in this chapter is not exhaustive, and other important factors may need to be addressed and optimized depending on the bioassay system.

6 Development of Thaw-and-Use Cells

While many bioassays use a Master Cell Bank (MCB) and Working Cell Bank (WCB) for continuous production, thaw-and-use (T&U) cells (also called ready-to-use or assay-ready cells) can be used for a bioassay straight out of the vial [2, 3]. These cells offer convenience and reproducibility and are rapidly being adopted for cell-based bioassays. Through manufacturing process development and optimization, T&U cells can be produced in large quantities with each batch harvested from the same working cell bank at defined scale-up conditions. Assay variations caused by daily cell culture are eliminated through a controlled process. Bulk production and storage of cell banks save on labor and time, reduce long-term development costs, and facilitate assay transfer between different laboratory sites. The simple and homogeneous nature of the reporter bioassay format goes hand-in-hand with more easily adoptable T&U cells.

This section will discuss general considerations in producing T&U cells for reporter bioassays. In particular, critical factors affecting cell culture, cell freezing, and long-term storage in the context of functional testing parameters will be examined.

Cell culture conditions prior to harvesting for T&U cells are critical for bioassay performance. Different culture vessels can impact cell quality and assay performance by affecting cell growth characteristics (cell growth rate and viability) and protein expression level (Fig. 6). T-flasks are standard vessels during the early stage of bioassay development; for large-scale cell production, many choices of culture vessels are available to produce several liters of culture. Testing a variety of culture vessels, such as flasks, spinners, roller bottles or bioreactors, is important for developing a process for scale-up. These vessels should be tested and chosen depending on the cell types (adherent or suspension), desired batch sizes, and other practical factors such as ease of handling, yield, and production cost. During cell expansion, cell seeding density, culture volume per vessel, and cell passing schedule need to be standardized to ensure batch-to-batch consistency. The cell doubling rate and viability should be recorded at each passing to

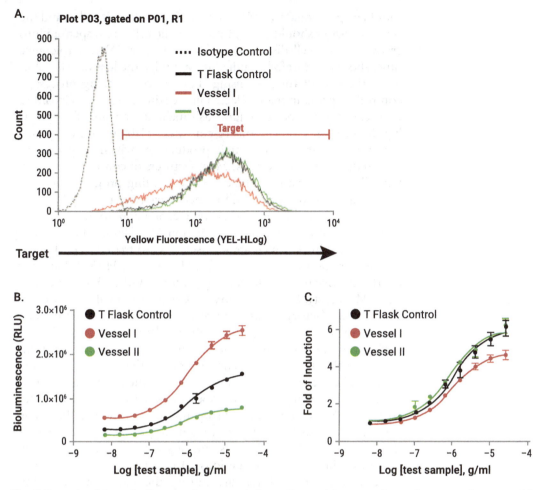

Fig. 6 Example of the selection of culture vessels impacting protein expression and assay performance. (**a**) FACS analysis showing different expression level for a cell surface target of a biologic product, from the cells produced by T flask, culture vessel I, and vessel II. (**b, c**) Functional bioassay showing different assay performance from the cells produced by T flask, vessel I, and vessel II. The cells produced from culture vessel I showed lower target expression level (**a**), higher basal luciferase activity and RLUs in the absence of drug stimulation, (**b**) and lower drug-induced assay window (**c**) when compared with the cells from T flask control and vessel II

monitor cell growth and help troubleshoot in the event of unexpectedly low production yield and undesired assay performance.

Cell-freezing conditions are another critical factor impacting T&U cell performance. Freezing conditions for T&U cells can be very different from WCB and MCB because T&U cells will be directly seeded into microplates for biological potency testing. Cell-freezing media can be optimized for the concentrations of FBS and cryoprotectant or choices of cryoprotectants. Many serum-free freezing medium products are commercially available but need to be carefully evaluated to make sure they do not impact cell performance. At the time of cell harvest, freezing medium should be precooled to 4 °C before adding to the cell pellet, and the cell suspension should be kept on ice while cells are dispensed into cryovials. The cell-dispensing time course, or DMSO tolerance time, should be carefully studied to minimize the loss of cell viability at the end of the dispensing. For the cell-freezing process, a controlled, programmable electronic freezing unit (e.g., CryoMed Freezer) with the ability to rigorously maintain the rate of cooling is highly recommended to ensure consistent viability and assay performance for the whole batch production, which may contain hundreds to a thousand vials. Manufacturers of controlled freezers typically recommend cell type-specific freezing programs though further modifications may still be necessary after the cell lines are engineered with exogenous protein expression to improve the consistency of cell quality and functionality. Unlike the majority of tumor cell lines, which are very sensitive to freezing conditions, certain common cell lines such as CHO-K1 and HEK293 are not overly sensitive to the cell-freezing protocol. Other freezing chambers (Mr. Frosty, Styrofoam box, or CoolCell) can also be used to produce satisfactory results. However, they are not recommended in a manufacturing setting for producing large quantities of T&U Cells due to the possibility of introducing large intrabatch variation.

Once the T&U cell batches are made, they should be transferred and stored in the vapor phase of liquid nitrogen or a $-140\ °C$ freezer. Functional performance was compared in multiple cell lines stored at $-80\ °C$ or in liquid nitrogen. The data showed that some cell lines, such as CHO-K1, can maintain good cell viability after 2–3 weeks in $-80\ °C$ freezers, although the viability might still change during longer time periods. Other cell lines, such as Jurkat T cells, lost 20% cell viability during the same time duration (Table 1). In general, we highly recommend storing the T&U cells at liquid nitrogen temperature to maintain consistent assay performance for the long term.

Table 1
Change of cell viability for thaw-and-use cells upon short-term storage at $-80\ °C$

		Postthaw cell viability			
	Days	Jurkat, %	Raji, %	CHO, %	HEK293, %
Stored at $-80\ °C$	0	86	91	99	82
	3	85	91	97	89
	7	77	92	98	89
	10	72	91	95	87
	14	69	91	94	87
	17	65	91	93	85

T&U cells produced with Jurkat, Raji, CHO-K1, or HEK293 cell lines were transferred from liquid nitrogen storage to $-80\ °C$ freezers and stored for the time period as indicated, up to 17 days. The cells were transferred back to liquid nitrogen storage after each time point until all test conditions were complete. The cells were then thawed, and postthaw viability was measured by trypan blue staining

7 Bioassay Prequalification

At this point, a bioassay must be further qualified for potency testing of drug products by demonstrating the accuracy, precision, linearity, and range following a series of repeated assays. Classic prequalification outcomes and metrics include establishing that the reporter bioassays can be performed using cells from continuous cell culture or in cryopreserved T&U format. If cells are used from continuous cell culture, their staging is critical for a consistent assay response. The previous section on assay optimization details some of the factors affecting cell functional responses.

For relative potency testing, T&U cells provide an important advantage over continuous culture cells as they repeatedly yield a consistent response across the batch of cells. This consistent response makes prequalification development far simpler as it eliminates the variability often associated with staging continuous culture cells. Large numbers of T&U cell vials can be prepared and stored below -140 °C, ensuring consistent responses for many years.

Accuracy and precision among replicates are important for any bioassay, especially for successful potency testing. Replicate precision will allow for better discernment of small potency differences and overall absolute reproducibility (e.g., EC_{50} and response to a standard concentration range during the test). Poor replicates with %CV > 5% can be an indication of some protocol or assay variable that was not fully optimized. Following good pipetting practices, including the choice and condition of equipment, can play a major role in bioassay success. It is undeniable that, for cell dispensing and serial dilution purposes, quality data depend on pipettes in good operating shape and without any biases they may introduce during dispensing. Choose the proper pipette for the working volume needed and ensure all channels of a multichannel pipette are in good working order without internal piston leakage. For dispensing purposes, electronic pipettes provide rapid and accurate dispensing, and facilitate larger and complicated experiments.

Begin by establishing and understanding the ligand or biologic product response as previously discussed in the section on assay optimization. Protocol optimization to generate a consistent EC_{50} is important as this impacts the EC_{80} and EC_{20} results. Agonist EC_{20} and EC_{80} are especially critical if used in blocking bioassays. The appropriate agonist concentration at either EC_{80} or EC_{20} depends on how big the assay window is. An agonist EC_{80} will be chosen if the assay has a fold induction less than a few hundred. When fold induction reaches over a thousand for certain bioassays, the assay will suffer from "over-sensitivity" and the potency assay will tend to fail during statistical analysis due to the large variation in RLU at the upper asymptote. In this case, the EC_{20} will be selected instead to reduce assay variation while still maintaining assay robustness.

Establishing the dilution series of the ligand or effector molecule is an important first step for potency assay development and prequalification. Identifying any potential biases or location effects across the entire plate, including rows, edges, and corners, will dictate which wells can be reliably used for samples. Often, the peripheral outside rows and columns of a 96-well plate can be prone to bias, but this issue should be empirically tested and not assumed. With 96-well plates, using the inner 60 wells for samples is a safe starting point until the outer rows and columns can be demonstrated to lack bias. Often, plate location bias can be remedied by changes in the protocol, such as preplating cells for a minimum of 1 h prior to sample addition. With the identification of which wells can be used for samples, the number of points for each dilution series can be established.

Sample placement and location throughout the plate is dependent on the number of series and data points for each series and any additional controls that might be used as part of assay acceptance. Place samples throughout the useable area of the plate in a non-clustered fashion to minimize any unintended response bias, including any potential luminometer plate reader bias. Alternating rows or columns is usually sufficient.

Potency assays typically encompass full dose responses of the test, reference, and control biologic samples. It is imperative to choose a dilution range such that a full response is created. The responses are then fitted using curve-fitting software, often as a four-parameter curve. Any potency difference is noted as a left or right shift across the x-axis if the curves are determined to be parallel either by an F-test or equivalency test. Each fitted curve should have an adequate number of data points to accurately establish the upper and lower asymptotes and as many points as possible on the linear range of the response curve containing the inflection point. Generally, the starting concentration of the test sample is serially diluted (e.g., three- to fivefold) before adding it to the cells, but other dilution schemes are possible. As shown in Fig. 7, the dilution series chosen will impact where the data points end up being located across the bioassay response. The dilution factor, starting concentration of sample, and number of data points in the series can all be manipulated to achieve a full response for the samples. Ultimately, this sample concentration range will need to accommodate samples with a potential change in potency with a response shift relative to the reference. For the purposes of assay prequalification, potency samples can be prepared by intentional dilution to create mock potency samples representing 50%, 70%, 140%, and 200% of the reference. After a series of repeated tests, this potency range and the resulting recoveries are used to establish the linearity response of the potency assay.

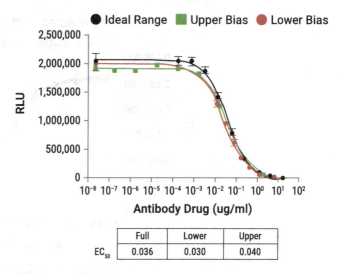

Fig. 7 Dilution series of an agonist-blocking antibody can impact curve fit. Agonist blocking by an antibody drug was demonstrated, starting at 20 μg/ml, using a series of data points across the entire response range (1:4.25-fold in black), one series with a bias of points at the lower asymptote (1:2.5-fold in red), and a third series (1:10-fold in green) with a bias at the upper asymptote

Special considerations should be noted for designing blocking bioassays, where a single concentration of ligand, typically the EC_{80} response, is inhibited by a titration of blocking antibody. It is important to understand if the antibody reacts to a surface receptor on the reporter cell or the ligand itself. Protocol adjustment should reflect the antibody's target: for a surface receptor; the antibody sample may need to be added first and preincubated with cells prior to addition of the ligand. If the antibody targets the ligand itself, the antibody and ligand should be coincubated for some amount of time before addition to the reporter cell as a sample. To create a robust protocol, this preincubation time should be empirically determined with a time course experiment as shown in Fig. 8.

8 Qualifying Potency Bioassays

Once the assay optimization and prequalification phase is completed, the assay moves into the qualification phase. During this phase, the assay design is confirmed to be capable of generating reproducible results for the specified purpose. Therefore, for a reporter gene bioassay, the assay qualification can be defined as a set of experiments performed under defined assay conditions, aimed to demonstrate that the method is capable of reliably measuring the relative potency (RP) of the drug under investigation. For best practices, the assay qualification should be conducted following a preapproved test protocol generated during the

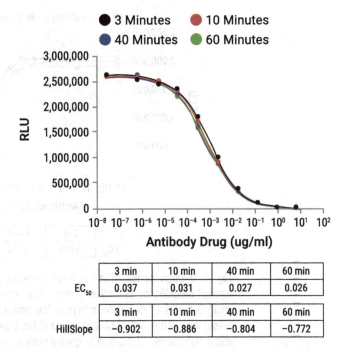

Fig. 8 Time course of antibody and ligand. A time course for the preincubation of an agonist with antibody drug demonstrates a response bias at short incubations where equilibrium has not been reached between the antibody and its agonist

prequalification phase that outlines the assay conditions, plate layout, number of sample replicates, and experimental design for each parameter investigated.

The following parameters are typically assessed during a reporter bioassay qualification: sample and assay suitability criteria, specificity, repeatability, intermediate precision, accuracy, dilution linearity, and range.

At the completion of the assay qualification, a report is generated that summarizes results, analysis, and conclusions on the assay achieving its intended purpose.

8.1 Bioassay Method and System Suitability Acceptance Criteria

A defined assay procedure is generated at the conclusion of the assay optimization/prequalification and prior to the assay qualification. A written method that includes the drug serial dilution, plate layout, number of samples and standard replicates on each plate, reagent concentrations, and incubation time is generated and made available to all the scientists performing the assay qualification. Typically, for a reporter gene bioassay, the plate layout will include multiple replicates of the drug sample and the drug standard, which are then used to calculate the relative potency of the drug sample. The relative potency of the drug is typically calculated as the EC_{50} ratio of the sample and standard curves [4, 5].

In addition to the assay conditions, the method should include a series of system suitability and sample acceptance criteria. Each plate should be evaluated against these criteria prior to proceeding with evaluating the qualification parameters. Doing so will ensure that only valid assays are included in the qualification analysis. The system and sample acceptance criteria should be derived from the assay prequalification data and usually will include a measure of how well each standard and sample curve fits, a measure of the replicates agreement, a minimum signal-to-noise ratio (or A/D ratio) for each sample and standard curve, and a measure of parallelism between the standard and the sample curve. Parallelism between the sample and standard curve can be assessed by several statistical analysis models [4, 5].

During assay qualification, the relative potency of the sample is calculated and reported for each plate that meets all the assay and sample acceptance criteria. For each plate that does not meet the assay and sample acceptance criteria, the plate is invalidated and repeated. If a high incidence of failed assay/samples acceptance criteria is observed during the assay qualification, steps should be taken to investigate the failed results and address the cause. If necessary, additional assay optimization or prequalification experiments may be conducted to ensure the consistency of the assay performance across different days and as performed by different scientists.

8.1.1 Qualification Parameters

The following parameters are typically evaluated for a reporter bioassay:

1. Specificity.
2. Precision.
   (a) Repeatability.
   (b) Intermediate precision.
3. Relative accuracy.
4. Dilutional linearity.
5. Range.

A definition of each parameter for general analytical procedure can be found in the ICH Harmonized Tripartite Guideline [6, 7].

Specificity is the ability of the bioassay method to specifically detect the potency of a drug. The bioassay should be specific to the receptor or signal pathway of the drug under investigation. Specificity can be tested using formulation buffer prepared as a sample to ensure the noninterference of the buffer with the bioassay. Additionally, the specificity can be tested using a sample drug not specific to the targeted pathway. In both cases, no dose–response curve should be observed when compared to the standard curve tested on the same plate.

Precision of a method expresses the closeness of agreement between a series of measurements obtained from multiple testing of the same sample. Precision of the bioassay is considered at two levels: repeatability and intermediate precision. Repeatability (also defined as intraassay precision) expresses the precision under the same operating conditions over a short interval of time. The precision of the method is expressed as the %CV of the series of measurements.

In a reporter gene bioassay, repeatability can be tested over one assay setup by one analyst. Sample and standard preparations are loaded into multiple assays plates in one experimental setup according to the method. If available, the drug product can be used as sample; alternatively, reference standard material can be used to prepare a "mock" sample. The sample and standard dilutions should be prepared independently. The number of plates included in the single setup will depend on the complexity of the method (4–6 plates are recommended). Each assay plate will generate a single reportable percent relative potency value and the intraassay precision will be reported as the %CV or % geometric coefficient of variation (%GCV) of all the reportable percent relative potency values.

Intermediate precision (also defined as interassay precision) expresses within-laboratory variations such as different days, different analysts, different equipment, and different lots of bioassay reagents. In a reporter bioassay, the intermediate precision experiments are performed by at least two analysts in multiple independent assay setups and on different days. Similar to repeatability, the drug product or reference standard can be used as a sample. Each analyst will prepare all assay reagents and samples independently and will use a different luminometer to analyze the plates. A different lot of thawed cells can be used by Analyst 2 in this portion of the qualification. Each assay plate (8–12 plates total is recommended) will generate a single reportable percent relative potency value and the interassay precision will be reported as the %CV or %GCV of all the reportable percent relative potency values.

The accuracy of an analytical procedure expresses the closeness of agreement between an accepted reference value and the value found. The accuracy of the reporter bioassay can be measured using a reference standard to prepare samples at different concentrations relative to the nominal drug concentration. Typically, samples at 50%, 75%, 100%, 125%, and 150% are prepared and tested against the reference standard material prepared at 100% of the nominal drug concentration. It is also not uncommon to dilute samples at 50%, 70%, 140%, and 200%. The individual and mean relative potency values for each sample concentration are reported. The individual and mean % biases (from 100% nominal value) are also calculated and reported.

The linearity of an analytical procedure is its ability (within a given range) to obtain test results that are directly proportional to the concentration of the sample. For the reporter bioassay, the linearity is calculated by plotting the experimental log relative potency values for each concentration versus the theoretical log relative potency values on a linear scale and by performing a linear regression analysis. The coefficient of determination (R^2), slope, and *y*-intercept from the linear regression analysis are calculated and reported.

The range of an analytical procedure is the interval between the upper and lower sample concentration for which it has been demonstrated that the analytical procedure has a suitable level of precision, accuracy, and linearity. The analysis and conclusions derived from the assessment of the repeatability, intermediate precision, relative accuracy, and dilutional linearity are used to establish the bioassay range over which results can be reliably reported.

8.2 Qualification Report

Subsequent to the execution of the protocol, all qualification results, data analysis, and conclusions are summarized in a report. It is recommended that information about critical equipment and reagents used (such as FBS and T&U cells) is also included.

The report should also contain a statement summarizing the qualification status of the bioassay method.

9 Challenges for IO Bioassay Development and Conclusion

Developing a cell-based functional assay to reflect the MOA of a drug that will be accessible at early stages of drug discovery will provide a smooth transition to support product lot release. Current methods to measure the potency of drugs for immunotherapy targets rely substantially on in vitro binding assays, primary T cell-based cytokine release assays, and in vivo model systems. Although in vitro binding assays satisfy high-throughput needs, lack of correlation with cellular functional response makes this method unreliable to screen out functional antibody candidates. Although antibody candidates can display high binding/blocking affinities, they may not display any functional response at the cellular level. Primary cell-based assays and in vivo model systems better reflect MOA but operate at lower throughput. High reliance on primary cells without continuing culture ability, as well as donor-to-donor variations, limit their ability for antibody screening in early drug discovery and lot release analysis. There is an urgent need for a simple, robust, plate-based functional bioassay to measure the potency of a drug candidate. Such an assay requires high sensitivity with appropriate specificity, precision, and accuracy for drug screening and characterization in early drug discovery, lot release, and stability studies. Luciferase reporter bioassays are designed based on cellular signal cascades responding to drug treatment. Activation of the corresponding pathway triggers luciferase gene transcription, and a luminescent signal can be read out using a luminometer. Due to its inherent sensitivity, large signal dynamics, and simplicity to set up, this reporter assay platform has been widely used in high-throughput screening for decades. The assay specificity is even more reliable for biologic development due to the cell surface nature of all targets. Some concerns for small-molecule screens, such as false positives or signaling events distal from receptor activation, are less applicable for large molecules. During bioassay development, cells are considered critical reagents and reproducibility is extremely important to qualify an assay for drug lot release. Further development of reporter-based functional assays into a T&U format eliminates the burdens of daily cell culture and variables introduced by cell health and cell preparation from other factors that could directly contribute to assay variability. Proper functional QC and optimized criteria for the number of cells per vial, cell viability, and mycoplasma and bacteria contamination testing will ensure assay consistency and ease of bioassay transfer from one location to another.

Designing a successful cell-based bioassay reflecting a drug's MOA requires a clear cellular mechanism, and many designs rely on publications demonstrating target validation in vivo with a well-studied antibody as positive control. Accessibility of the control antibodies is critical to validate bioassays. In many cases in immunooncology, the targets of interest have been recently identified and antibodies are either proprietary or inaccessible from any commercial sources. Collaborations with pharma–biotech companies or reputable academic labs are critical to validate the assay design. Even when the antibodies in publications can be accessed from commercial sources, most of those antibodies are purified for research use only. The formulation of the products, especially the presence of the preservative sodium azide, often produces a hook effect at high antibody concentration, creating major challenges for assay optimization. In addition, these research-grade antibodies are mostly qualified for cytokine release, flow cytometry, and Western blotting. Therefore, it can prove challenging to apply these antibodies to T&U cells to demonstrate the suitability of a bioassay for testing biological potency.

Many immunotherapy drug targets show promising responses in mouse models without a clear understanding of their cellular mechanism of action, which creates a hurdle for designing cell-based assays for antibody screening and assay validation. Some targets are clinically relevant but a clear understanding of their ligand or corresponding receptors is unavailable. B7-H4, for example, was discovered as a B7 family member molecule that is responsible for T cell immunity [8]. Ligation of T cells with B7-H4 has a profound inhibitory effect on T cell growth, cytokine secretion, and development of cytotoxicity. However, the T cell receptor responding to B7-H4 is still unknown as well as how the inhibition is mediated. In other cases, some targets are reported to have multiple ligands, and the clinical significance of each target–ligand interaction is largely still under investigation. The V-domain immunoglobulin suppressor of T cell activation (VISTA) is a negative immune-checkpoint protein that controls a broad spectrum of innate and adaptive immune responses [9, 10]. However, the ligand-or-receptor paradigm of VISTA in regulating T cell activation is unclear. VSIG3, VSIG8, and VISTA all interact with VISTA [11–13]. The immunoinhibitory molecule Lymphocyte-Activation Gene 3 (LAG-3, CD223) synergistically regulates T cell function with PD-1 to promote tumoral immune escape [14, 15]. Major histocompatibility complex Class II (MHC-II) is the canonical ligand for LAG-3, but it remains controversial whether MHC-II is solely responsible for the inhibitory function of LAG-3. It was reported recently that a newly identified ligand Fibrinogen-like Protein 1 (FGL1), a liver-secreted protein, is a major LAG-3 functional ligand independent from MHC-II [16]. FGL1 inhibits antigen-specific T cell activation, and ablation of FGL1 in mice promotes T cell immunity. Poor clinical outcomes from several MHC-II blocking anti-LAG-3 mAbs evaluated in clinical trials for the treatment of advanced human cancer may suggest that these antibodies do not block the clinically relevant ligand. Moreover, identifying and determining the availability of a biologically relevant cell line background to study the cellular pathway of a target of interest is critical and sometimes a limiting factor in designing a valid cell-based assay.

Immunotherapy is a novel, rapidly evolving cancer treatment with exciting benefits, but it also presents unique challenges for validating targets and determining their clinical roles in cancer treatment. With many immunotherapy trials in clinics, what is known today is very likely to change tomorrow. Developing cell-based functional bioassays that reflect the true MOA in this dynamic area, and embracing the challenges, will guide future research in this rapidly growing field.

References

1. Robinson CJ, Sadick M, Deming S et al (2014) Assay acceptance criteria for multiwell-plate-based biological potency assays. Biopharmaceutical Best Practices Association, Citrus Heights, CA
2. Gazzano-Santoro H, Chan LG, Ballard MS et al (2014) Ready-to-use cryopreserved primary cells: a novel solution for QC lot release potency assays. BioProcess Intl 12:28–39
3. Cheng ZJ, Garvin D, Paguio A et al (2014) Development of a robust reporter-based ADCC assay with frozen, thaw-and-use cells to measure Fc effector function of therapeutic antibodies. J Immunol Methods 414:69–81
4. USP 35-NF 30 (2012) USP Chapter <1032> Design and development of biological assays. USP Pharmacopeial Convention

5. USP 35-NF 30 (2012) USP Chapter <1034> Analysis of biological assays. USP Pharmacopeial Convention
6. Guideline, ICH harmonised tripartite. Validation of analytical procedure: text and methodology Q2 (R1)
7. USP 35-NF 30 (2012) USP Chapter <1033> Biological assay validation. USP Pharmacopeial Convention
8. Sica GL, Choi IH, Zhu G et al (2003) B7-H4, a molecule of the B7 family, negatively regulates T cell immunity. Immunity 18:849–861
9. Yapa EH, Roschea T, Almob S et al (2014) Functional clustering of immunoglobulin superfamily proteins with protein-protein interaction information calibrated Hidden Markov model sequence-profiles. J Mol Biol 426(4):945–961
10. Flies DB, Han X, Higuchi T et al (2014) Coinhibitory receptor PD-1H preferentially suppresses CD4+ T cell-mediated immunity. J Clin Investig 124(5):1966–1975
11. Xu W, Hiếu TM, Malarkannan S et al (2018) The structure, expression, and multifaceted role of immune-checkpoint protein VISTA as a critical regulator of anti-tumor immunity, autoimmunity, and inflammation. Cell Mol Immunol 15:438–446
12. Wang J, Wu G, Manick B et al (2019) VSIG-3 as a ligand of VISTA inhibits human T-cell function. Immunology 156(1):74–85
13. Wang J, Manick B, Renelt M et al (2018) Poster: VSIG-8 is a co-inhibitory ligand and an immune checkpoint molecule for human T cells. J Immunol 47(4)
14. Woo SR, Turnis M, Goldberg M et al (2012) Regulate T-cell function to promote tumoral immune escape. Cancer Res 72:917–927
15. Nguyen LT, Ohashi PS (2015) Clinical blockade of PD1 and LAG3—potential mechanisms of action. Nat Rev Immunol 15:45–56
16. Wang J, Sanmamed M, Datar I et al (2019) Fibrogen-like protein 1 is a major immune inhibitory ligand of LAG-3. Cell 176:334–347

Open Access This chapter is licensed under the terms of the Creative Commons Attribution 4.0 International License (http://creativecommons.org/licenses/by/4.0/), which permits use, sharing, adaptation, distribution and reproduction in any medium or format, as long as you give appropriate credit to the original author(s) and the source, provide a link to the Creative Commons license and indicate if changes were made.

The images or other third party material in this chapter are included in the chapter's Creative Commons license, unless indicated otherwise in a credit line to the material. If material is not included in the chapter's Creative Commons license and your intended use is not permitted by statutory regulation or exceeds the permitted use, you will need to obtain permission directly from the copyright holder.

Chapter 10

Miniaturized Single Cell Imaging for Developing Immuno-Oncology Combinational Therapies

Thomas Jacob, Pavani Malla, and Tania Vu

Abstract

Targeted oncotherapeutics offer progress in improved treatment for individuals diagnosed with cancer; however, patient-developed resistance, both to conventional chemotherapies and to single agents, remains a major hurdle. A current approach to combating drug resistance and risk of cancer recurrence is to find combinations of drugs that effectively target multiple functional pathways, increasing the prospect of providing more durable cancer cell kill. However, combination drug treatment introduces a daunting number of options as the plethora of small molecule and antibody immunotherapy compounds available continues to grow, presenting major challenges in identifying candidate combination pairs of suitable, high effect. Here we describe a miniaturized single cell imaging approach for evaluating ex vivo combination drug responses in primary patient cells which uses minimal sample requirement, enabling screening of up to 50 drug conditions per one million primary cells. This approach also employs the capability to perform digitized molecular counting to quantify surface and intracellular protein targets within single cells that may often be present at low levels in single cells (e.g., phosphoproteins, immune modulatory molecules induced via inflammation). This miniaturized single cell imaging approach offers new capabilities for assessing the functional/pathway status of primary patient single cells from distinct cell subpopulations and profiling the therapeutic effect of multiple protein targets to combinations of drugs, with single cell, high-content information.

Key words Single cell, Miniaturized assay, Imaging, Drug screening, Drug combination, Drug response, Drug resistance, Immunotherapy, Leukemia, Cancer

1 Introduction

A major challenge in identifying combination drugs of high therapeutic potential is a lack of platforms available to investigators for functional, detailed assessment of numerous combination drug treatments on primary patient cells. Flow cytometry is traditionally-used option, however the requirements of nearly millions of patient cells for testing each drug condition makes it limited in practice for routine screening of increased number of drugs. Moreover, many protein targets (e.g., phosphoproteins, nuclear proteins, and immune modulatory molecules induced via

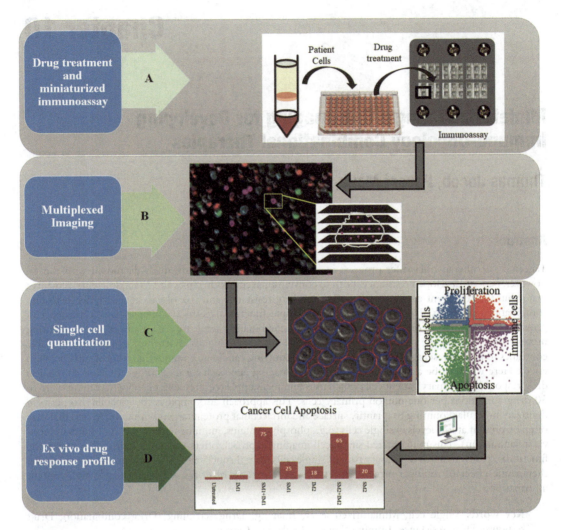

Fig. 1 Miniaturized single cell imaging workflow

inflammation) are often present at low levels, and thus often challenging to detect and quantify. In such scenario, negative detection cannot be conclusively interpreted, and it is difficult to discriminate the signal from diffuse background noise. Other ex vivo screening platforms such as colorimetric assays for cell viability or cell death (e.g., MTS/MTT assays) lack single cell granularity and hinder the detection of cell subpopulations that may be insensitive to drug treatments.

We describe below a new imaging assay approach that addresses these current limitations in quantitative and functional evaluation of primary patient cellular response to targeted combination drugs (Fig. 1). This assay platform is miniaturized in format to minimize the patient cell requirements and maximize the screening of drug conditions. Additional important technical capabilities include multiplexing of biomarkers for profiling cellular identity and pathway activity in single cells' subpopulation. Cells are treated with a panel of desired drugs as single agents and in combination (e.g., single immunotherapy, single small molecule, combination immunotherapy, and small molecule targeted drug) and labeled by fluorescent probes in a multi-well chamber and are identified by cell segmentation; and fluorescent intensity of every cell is quantified by fluorescence microscopy that employs diffuse fluorescence quantitation or as an option, the use of bright nanoparticles such as quantum dots to perform "digitized molecular counting" of discrete proteins [1]. This fluorescence quantitation is collected from each single cell and thus provides high granularity in simultaneous readouts of critical markers of functionality of cellular subpopulations of relevant clinical interest.

Below we provide methodology for assessing combination immunotherapy and small molecule targeted agents from leukemia patient blood samples. This technology provides new quantitative information for evaluating the specific physiological responses such as cell proliferation and cell death in myeloid cells and T-cells (Fig. 2). This information, which comprises of patient specific responses of every single cell to various drug conditions can be evaluated alongside existing other available information, such as high content flow cytometry of baseline patient status as well as genetic-mutational information for interpreting drug combinations of broad effect in individual cohorts. This methodology can also be extended to other cancer sample types and classes of drug combinations. Given the active areas of single cell imaging-based assays in drug discovery and development [2], the need to better select and test combination therapies for clinical trials from the growing number of drug candidates [3] and the active growth of particular classes of drugs (e.g., immunotherapy checkpoint inhibitors) [4], we trust that this miniaturized imaging approach will aid to accelerate understanding and nomination of effective drug combinations that can be advanced to clinical trial testing.

2 Materials

2.1 Cell Preparation

1. Patient blood.
2. Cell lines (e.g., MOLM14, Jurket) (*See* **Note 1**).
3. Iscove's Modified Dulbecco's Medium (IMDM), for diluting blood.
4. Ficoll-Paque PLUS density gradient media (GE healthcare Life Sciences).

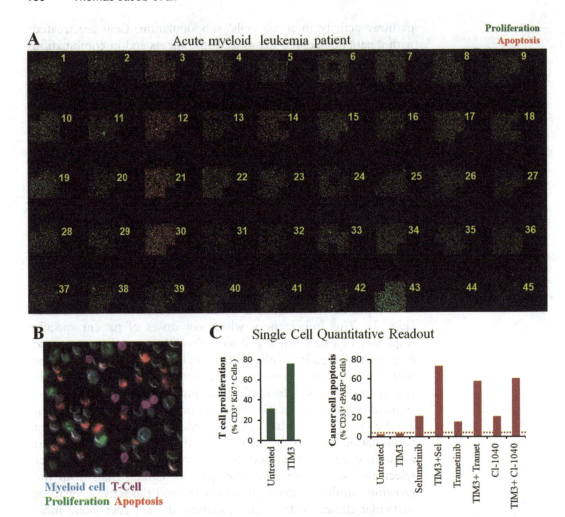

Fig. 2 Miniaturized single cell imaging readouts. (**a**) Example multi-well imaging data showing cell death (red) and proliferation (green) from a blood sample of an acute myeloid leukemia patient. Each drug well is comprised of 9 regions of interest (ROIs). Dark areas represent ROIs on the array that contain sporadic artifact (e.g. cellular clumps) and are not analyzed. (**b**) Higher resolution data from (**a**) showing simultaneous single cell readouts of myeloid cell, T-cell, proliferation and apoptosis. (**c**) Example of quantitative data from an AML patient showing stimulation of T-cell proliferation by TIM3 immunotherapy and combination TIM3 with particular MEK pathway targeted inhibitors showing synergistic myeloid cell kill that is enhanced compared to single agent MEK inhibitor alone

2.2 Cell Culture

1. 96-well culture plates.
2. Immunotherapy reagents, anti CD3 antibody, anti PD1 antibody, anti TIM3 antibody anti PDL1 antibody etc.
3. RPMI 1640 (culture media).
4. Fetal bovine serum (FBS).
5. 4% Paraformaldehyde in PBS (fixing solution).
6. PBS 1×.
7. Methanol.

2.3 Immunofluorescent Labeling (IF)

1. Assay chamber/Imaging (multi-well, glass bottom chamber) (*See* **Note 2**).
2. Fluorescent probe conjugated antibodies for labeling, organic dyes or quantum dot conjugates, respectively for diffuse or molecular digitized fluorescence quantitation.
3. Bovine serum albumin.
4. Normal goat serum.
5. Milli-Q H_2O.
6. PBS 1×.
7. DNA counter stain (Nuclear Green DCS1, Abcam).
8. Mounting solution (SlowFade antifade).
9. Coverglass sealing solution.

2.4 Image Acquisition

1. Fluorescent microscope (e.g. Carl Zeiss Axio Observer Z1, inverted microscope).
2. Motorized stage.
3. Objectives (e.g., 20× and 5× EC Plan-Neofluar 5×/0.16 WD = 18.5 M27, Plan-Apochromat 20×/0.8 WD = 0.55 M27).
4. Camera (e.g., ORCA Flash 4 LT, Digital CMOS, and Hamamatsu).
5. Light source SOLA SM 395 (Lumencor).
6. Image acquisition software (umanager, Zen (Carl Zeiss)).

2.5 Image Analysis

1. MATLAB software.
2. Cell profiler software (https://cellprofiler.org/).
3. R-Studio software (https://www.rstudio.com/).
4. Fiji ("Fiji Is Just ImageJ") (https://fiji.sc/).

3 Methods

3.1 Cell Preparation

1. Mononuclear cells (MNC) are purified from blood or bone marrow using standard Ficoll-Paque separation protocol according to the manufacturer's recommendations (GE healthcare Life Sciences). Briefly, 3 ml of patient blood or bone marrow is diluted with 3 ml of RPMI solution, and 6 ml of the diluted blood layered over 5 ml of Ficoll-Paque in a 15 ml tube, and spun using a swing out rotor at 400 × g for 15 min at 20 °C. MNC layer is retrieved and washed in IMDM solution.
2. Appropriate cell lines are used as positive control for protein biomarkers of interest in the labeling assays. For, e g.,

MOLM14 cells for myeloid cells and Jurket cells for T lymphocytes (T-cells) (*See* **Note 1**).

3.2 Cell Culture

1. MNC purified from the patient blood or bone marrow is directly used for the cell culture and drug treatments.
2. Cell culture is conducted in a 96-well plate. Cells were stimulated with anti-CD3 to increase T-cell proliferation. Coated wells with anti-human CD3 antibody at 1 μg/ml concentration, dissolved in PB, added 100 μl volume per well, and incubated for 4 h at 4 °C.
3. Control wells were also made by coating mouse IgG (isotype) instead of anti-CD3 antibody.
4. MNCs are prepared for culture by diluting in RPMI solution supplemented with 20% FBS (RPMI-20) at a density of 1 M cells/ml, and added 100 μl of the MNC solution (100,000 cells) per well.
5. Selected small molecule and immune-checkpoint inhibitor drugs are prepared at appropriate concentrations and added 100 μl/well. The final volume in each well is 200 μl, which consists of cells and inhibitor reagents.
6. Treat patient cells with the drug panel in standard culture conditions of 37 °C, 5% CO_2, for 3 days.
7. Cell lines used as the positive control were grown separately in RPMI-20 in standard culture conditions.
8. At the end of the cell culture and drug treatments, cells were fixed by adding PFA to 2% final concentration. Transfer to microcentrifuge tubes and spin to remove supernatant. Wash cells with PBS by centrifugation and store cells for use in media such as methanol until use.

3.3 Immuno Fluorescent Labeling (IF)

1. Rehydrate the stored cells preserved in methanol prior to cell staining with Milli-Q H_2O added to cells. Spin and remove supernatant.
2. Add ~10 μl of Milli-Q water to the cell pellet, titurate with a pipettor to mix. Transfer 15 μl of cells into a multi-well assay chamber with a detachable glass bottom surface. The working volume of such a well is 12–15 μl.
3. Add patient cells treated with different drug conditions in separate wells in the chamber. Also add standard cell lines in the control wells.
4. Briefly spin the chamber at low speed (90–100 × g), for 15 s, using a swing out rotor. Once the cells are settled on the surface, slowly remove the supernatant from each well using mild vacuum.

5. Add 15 μl of blocking solution (8% normal goat serum and 2% BSA) and incubate at RT for up to 1 h.

6. Prepare the multiplexed fluorescent probe solution by mixing the antibodies at optimal dilutions. These may be conventional dyes if employing standard diffuse fluorescence or quantum dot-conjugated nanoparticles if employing sensitive discrete quantum dot detection. Appropriate antibody dilutions must be optimized (*See* **Notes 3** and **4**).

7. Add antibody probe solutions and incubate for 2 h (1 h at 37 °C, and 1 h at RT). A few wells should be assigned as negative control that are assayed in the absence of antibody probes.

8. Remove the antibodies for the wells, separate the glass bottom, and wash the surface in PBS.

9. Prepare an appropriate nuclear dye (e.g., Nuclear Green DCS1 counter stain, Abcam) at optimal dilution, and label the cells for 5–10 min.

10. Wash the slides briefly in PBS after the counterstaining, mount the cells in compatible mounting solution, and seal the cover glass.

3.4 Image Acquisition

1. Images of the cells in each well position is acquired using a fluorescent microscope (Carl Zeiss Axio) equipped with a camera (Hamamatsu ORCA Flash 4 LT), motorized stage, optimal light source (SOLA SM 395, Lumencor), filter sets (Carl Zeiss), and appropriate acquisition software (e.g., umanager, Zen software (Carl Zeiss)).

2. Typically, the entire slide is first scanned at lower resolution (5×) in bright field. The area of each well is selected. Acquire images at bright field (DIC) and in multiple fluorescent channels corresponding to the labeled probes in the multiplex. For diffuse fluorescence detection of antibodies conjugated with conventional organic dyes, images are acquired at 20× magnification at ~3 slices with interdistance of ~2 μm. For discrete quantum dot for diffuse fluorescence detection of antibodies conjugated with conventional organic dyes, images are acquired at 20× magnification at ~3 slices with interdistance of ~2 μm. 40× magnification for discrete quantum dot probe detection.

3. Fluorescent channels are acquired from multiple *z*-planes appropriate to the size of the cells and intensity of the fluorescent probe. For MNCs, three slices with interdistance of ~2 μm are appropriate for conventional organic dyes and ten slices with interdistance of 50–100 nm are appropriate for QD nanoparticle detection. All images from a single well are named with a well number and image number (W1_1, W1_2 etc.) for

unique identification purpose. All images from one field are saved as a single stacked image file consisting of multiple channels and multiple slices.

3.5 Image Analysis

1. First, the cells are segmented from the acquired images (e.g., Cell Profiler software). Cell detection algorithms are customized for the single cell detection from the nuclear channel.

2. The total fluorescence from each cell is quantified (image J, custom algorithms in MATLAB). The stacked image files, along with segmented cell coordinates output file are processed by custom written MATLAB program, to extract the total fluorescence from each cell. The total fluorescence intensity for each cell is quantified in the selected channels and recorded. One CSV file output is generated for each channel, and in each well (*See* **Note 5**).

3. Cells that are positive for each marker is identified by a threshold selection algorithm using customized RStudio software. Single cell value CSV file output from the earlier step is used as the input data for threshold selection. Threshold is selected with respect to the positive and negative control wells in the given experiment.

4. A tabulated list of number of cells that are positive and negative for each marker is generated by the RStudio software. If the cells are not readily assigned as positive or negative, then the cells near the threshold line are assigned as ambiguous cell population. The number of cell percentages quantified for each marker is compared for each drug condition in a patient, and the function cellular response such as cell death and cell proliferation in each cell type (Myeloid, T-cell, etc.) is quantified.

5. Finally, the cellular response for relevant drug conditions for all the patients in the given study is compared. This provides the overall effect of a specific drug in given population of leukemia patients.

4 Notes

1. Cell line selection: Appropriate cell lines should be selected for each markers in the multiplexed assays. For, e.g., M 14 cell is positive for myeloid marker CD33. Jurket cells are positive for T-cell marker CD3. These cell lines are positive for cell proliferation marker Ki67, when untreated. Treating MOLM 14 cells with AC220 drug for 24–36 h at 1 µM concentration, for example, induces apoptosis, and a majority of the cells become positive for cell death marker cPARP.

2. Microwell assay chamber: Microwell chambers should have a glass bottom surface for high quality image acquisition. The bottom surface may be either glass slide or coverglass.

3. Fluorophore selection: Fluorophores selected for the antibodies should be compatible for multiplexed imaging. These fluorophores should not overlap in its excitation/emission spectral properties. For, e.g., Alexa 488, Alexa 555, and Alexa 647 are compatible fluorophores, when used with appropriate filter sets. DNA counter stain for nuclear labeling should be also compatible with the rest of the dyes. For, e.g., Nuclear Green DCS1 dye can be used in place of A488. Nanoparticle quantum dots offer high stokes shift and fewer issues of spectral overlap [5].

4. The reagent optimization: The optimal concentration of fluorescent antibodies and the nuclear counter stain is depending upon the assay conditions. The dilution of these reagents should be optimized for the assays in each laboratory condition. For, e.g., Light source and camera type in the microscope can have different levels of sensitivity levels and influence the effective signal to noise ratio of the fluorescence intensity in the cells.

5. Image acquisition and quality assessment: High throughput imaging will result in many hundreds of image files. It is possible that certain images may have artifacts originating from debris, air bubbles, or other particles. It is suggested to visually assess the images using Fiji or other similar software, to maintain the quality of acquired images. Images that has unacceptable levels of artifacts may be omitted or affected cells may be selectively removed from the analysis.

References

1. Jacob T, Argawal A, Rummuno-Johnson D, Tyner J, O'Hare T, Gonen M, Druker BJ, Vu TQ (2016) Ultrasensitive quantum dot phosphoassay system for single leukemia cell resistance detection. Sci Rep 6:28163

2. Pomerantz AK, Sari-Sarraf F, Grove KJ, Pedro L, Rudewicz PJ, Fathman JW, Krucker T (2019) Enabling drug discovery and development through single-cell imaging. Expert Opin Drug Discov 14(2):115–125. https://doi.org/10.1080/17460441.2019.1559147

3. (2017) Rationalizing combination therapies. Nat Med 23:113. doi:https://doi.org/10.1038/nm.4426

4. Topalian SL, Weiner GJ, Pardoll DM (2011) Cancer immunotherapy comes of age. J Clin Oncol 29(36):4828–4836. https://doi.org/10.1200/JCO.2011.38.0899. Epub 2011 Oct 31. PubMed PMID: 22042955; PubMed Central PMCID: PMC3255990

5. Vu TQ, Lam WY, Hatch E, Lidke D (2015) Quantum dots for quantitative imaging: from single molecules to tissue. Cell and Tissue Research 360(1):71–86. PMID25620410

Chapter 11

Precision Medicine: The Function of Receptor Occupancy in Drug Development

Leanne Flye-Blakemore, Christèle Gonneau, Nithianandan Selliah, Ajay Grover, Sriram Ramanan, Alan Lackey, and Yoav Peretz

Abstract

Over the last two decades, flow cytometry has increasingly been utilized during both early and late phases of drug development. Instruments are now able to simultaneously measure over 50 parameters at the single-cell level and within complex mixtures of cells with significant precision. Along with immunophenotyping and functional testing used for clinical immune monitoring, receptor occupancy (RO) assays have become a standard application in flow cytometry in the context of drug development, particularly in indications where biologics are the therapeutic modality. In this chapter, we aim to discuss the various assay formats of receptor occupancy and outline some of the advantages and disadvantages related to specific assay designs with potential technical and analytical pitfalls.

Key words Receptor occupancy, Target engagement, Biologics, Precision medicine, Pharmacodynamic, Flow cytometry

1 Introduction

Receptor occupancy assays (RO) using flow cytometry are robust, analytical methods designed to identify, quantitate, and monitor the binding of a therapeutic antibody (drug) to its cellular target (specific receptor) in specific cellular subsets. Also referred to as target engagement assays, they are often used to generate pharmacodynamic (PD) biomarker data and are coupled to assays for both anti-drug antibody detection (ADA) and traditional pharmacokinetic (PK) assessments for modeling PK/PD/ADA relationships. The assays are most typically used for exploratory endpoints but may include custom quality control reagents to meet primary and secondary endpoint validation requirements. The determination of RO can be obtained using different approaches and assay formats that can generate the following measurements or a combination thereof: free receptors, total receptors, bound receptors, receptor modulation [1, 2].

During preclinical phases, RO testing assists in compound screening, demonstration of target engagement, and determination of the minimal effective dose for human clinical trials [1, 3]. For example, the clinical study with TGN1412, a CD28 targeting agent, exemplifies the importance of the use of RO assay in dose selection for first in-human studies [3, 4]. Due to different pharmacological properties of TGN1412 between the nonhuman primates and humans, the no adverse effect level (NOAEL) determined in the preclinical toxicity studies was not appropriate to humans. The use of TGN1412 in the clinical trial resulted in the life-threatening cytokine release syndrome during safety testing [3, 4]. Had RO been assessed in the nonclinical safety studies and the 10% RO for minimum anticipated biological effect level (MABEL) approach used, the starting dose in human volunteers would have been 30,000 times lower [3–5]. During phase I clinical trials, RO assays are mainly used to confirm target occupancy observed in animal models. During phase II and phase III clinical trials, RO assays enable PK/PD modeling and help guide the dose of administered drug [1, 6, 7] to help maximize the desired clinical effect while minimizing secondary toxicities and adverse events. For example, anti-PD1, anti-PD-L1, AMG479, ATR107, Etrolizumab or VX15/2503 are among the therapeutic antibodies for which RO information on circulating cells has been used as PD biomarker [3, 8–10].

There are multiple technical and logistical challenges associated with the development, optimization, validation, and implementation of RO assays for multi-centric clinical trials. In addition, like any other flow cytometry assay used for multi-centric studies, the quality and longitudinal comparability of the generated data largely relies on optimal assay development and validation. Validation typically includes the selection of a robust quality control specimen, titration of detection reagents, specificity, sensitivity, target prevalence, specimen stability (pre- and post-processing), inter-instrument standardization processes, global standard operating procedures (SOPs), and appropriate staff training [11]. The design and implementation of these assays can follow several paths. All methods rely on a thorough assessment of the affinity of clones utilized for total and free receptor, the manner in which the background for each is determined, the stability of the therapeutic and K_d thereof in conjunction with sample stability, the response of target receptor once bound by drug (internalization, shedding), the presence of ADA and effects on RO measurements and, the total target expression levels coupled with the intensity of detection reagents ex vivo. In addition, a thorough knowledge of the therapeutic mechanism of action, structure of the drug and function is required for adequate assay design and implementation.

With the development of novel bispecific (BsAbs) and trispecific (TsAbs) therapies that recognize two or three unique epitopes, more complex assay design is required to ensure adequate assessment of the characteristics aforementioned. While multi-specificity opens these therapeutics to a wider range of applications such as redirecting T cells to tumor while blocking anti-tumor signaling pathways, or shutting down multiple signaling pathways simultaneously, or even delivering payloads to targeted cells [12], these therapeutics do require higher levels of scrutiny to ensure the assays for occupancy perform as intended. In addition, new small engineered antibody fragments such as diabodies (50 kDa non-covalent dimers of single-chain Fv fragments) are useful alternatives to their larger antibody counterparts but require unique modifications in order to detect them in traditional RO assays.

No matter the nature of the specificity of the therapeutic, the assay design can follow a specific workflow as described below for each possible format of RO, of which there are three main approaches: (1) free site receptor expression or (2) bound receptor expression combined or not with total receptor expression measurements, and (3) functional readout in response to receptor occupancy. The workflow involves competitive/non-competitive clone screens, proper background determination method, titrations of competitive, non-competitive and secondary detection reagents, RO saturation curve specific to the mechanism of action (MOA) of the drug, and sample stability in presence of compound.

In this chapter, we discuss the different technical approaches that can be used to assess RO by flow cytometry. In addition, we also explore the various advantages and disadvantages of each approach. Finally, we review the challenges that can be encountered during RO assay development, validation and implementation for multi-centric clinical trials. This chapter is not intended to be an exhaustive list of all possible methodologies that could be employed for RO, but will deliver an overview of basic methods, with pros and cons of each, so that the investigator can have a broad knowledge base with which to design their own drug-specific methods given special consideration for the mechanism of action of their unique compound. In addition, we will discuss special considerations to be taken into account before any RO assay is designed and validated for use in clinical trials.

2 Special Considerations for Low Expressing Targets (PD-1, OX-40, LAG-3)

There are special considerations that must be taken into account when investigating occupancy of receptors of lower copy numbers. Low-density receptor occupancy can be detected, but the dynamic range may be insufficient for assessing longitudinal RO. This can be countered by using signal amplification with secondary or even tertiary detection reagents bound to the drug or using a very high-affinity secondary detection reagent that is directly conjugated to a fluorochrome with the highest possible stain index (brighter fluorescence). In addition, if blood collection volumes are not limited, a larger starting volume of sample may be required to ensure a sufficient number of cells are analyzed that express the receptor of interest. However, before signal amplification methods can be determined, clone screening and initial titrations must be performed. A necessary step in assay development for low copy number receptors is to ensure accurate clone screening and detection reagent titrations by creating a mock sample in which expression is detectable. Many antibody therapeutics are targeted to activated or exhausted T cell phenotypes, or to tumor suppressor receptors which are expressed at minimal levels within the peripheral blood subsets. If no cell line is available with constitutive upregulation of these receptors, then upregulation in donor blood can be achieved by stimulating isolated peripheral blood mononuclear cells (PBMCs) using phytohemagglutinin-L, Concanavalin A, or similar stimulants. The goal of the stimulation is to increase expression to detectable levels without overstimulating such that cells are 100% positive. Overstimulation can result in decreased phenotypic markers like CD3 and CD4, making gating difficult. Because stimulation with lectins results in non-specific T-cell response, the process can also falsely increase background fluorescence. With this in mind, determination of correct titration calls and background staining cannot be determined on stimulated samples alone. It is also for this reason that stimulation protocols during clinical sample testing for RO are not recommended. However, stimulated sample testing is very helpful in clone screening for competition with the therapeutic.

Stimulated cells are either tested as plain PBMC or are spiked back into matched donor whole blood to perform additional testing. For the purpose of the methods described in this chapter, it will be assumed that the operator knows how to isolate PBMCs utilizing a density gradient media such as Ficoll®-Paque PREMIUM 1.073 (Sigma Catalog No. GE17-5446-52), or a similar product. The separated cells are then washed and resuspended in sterile RPMI + 10% FBS complete culture media at a concentration of 0.5×10^6 cells/mL media. Only the PHA-L stimulation process will be described. Methods utilizing other stimulants are similar and differ mainly with respect to the concentration of stimulant used.

2.1 Materials: PBMC Stimulation Using PHA-L

1. Centrifuge—Thermo Electron Corporation, Model # IEC CENTRA GP8 (or equivalent).
2. Revco CO_2 Incubator, Model # RCO3000TABA (or equivalent).
3. Laminar Fume Hood—Baker Class IIA, Model # 56-600 (or equivalent).

4. Hematology Analyzer, Sysmex, Model # XS1000i (or equivalent).
5. Rainin pipettes, 1 μL to 10 mL (or equivalent).
6. 12 × 75 Polypropylene Test Tubes—Beckman Coulter, Part # 2523749 (or equivalent).
7. 12 mm Flange Plug Caps—Globe Scientific, Product # 118127R (or equivalent).
8. 15 mL Conical Centrifuge Tubes—Sarstedt, Cat # 584.002 (or equivalent).
9. 48 Flat Well Tissue Culture Plates—Fisher, Cat # 08–772-1C (or equivalent).
10. Leukoagglutinin PHA-L, Cell Culture Tested, Lyophilized—Krackler Scientific, Cat # 45-L4144-5MG-EA (or equivalent).
11. RPMI with L-Glutamine 1640 (1×) Liquid, Gibco® Invitrogen Cell Culture, Cat # 11875-093 (or equivalent).
12. 10% FBS Cell Culture Media.

2.2 Method: PBMC Stimulation Using PHA-L

1. Deliver 1 mL of sterile PBMC cell suspension (0.5×10^6 cells/mL in complete culture media) to each of 12 wells in a sterile 48-well plate.
2. Add 10 μL of a 1 mg/mL solution of PHA-L into the first 6 of 12 wells. Mix gently with sterile pipet. Deliver nothing to the other six wells. These will be the unstimulated controls.
3. Place the plate into a 37 °C, 5% CO_2 humidified incubator for 24 h. This is typically enough time to upregulate the necessary receptors without overstimulating. If testing indicates that the receptor is not upregulated, repeat the process and incubate for 48 h.
4. Remove all PHA-L stimulated cells from the first six wells and combine into 1–15 mL conical tube. Place the combined unstimulated wells into a separate 15 mL conical tube.
5. Centrifuge cells at 400 RCF for 5 min, decant and resuspend in 6 mL of ambient stain buffer.
6. Obtain a count and adjust concentration of cells to 10×10^6 cells/mL using stain buffer.
7. Proceed with RO assay methodology on either the plain PBMCs, or spike the PBMCs into fresh, donor-matched whole blood at approximately 20–30% based on the WBC of the whole blood sample. Proceed with RO assay method of choice.

3 Special Considerations for Detection Reagents

3.1 Detection Antibody: Clone Screening

After determining the mechanism of action of the therapeutic and selecting the ideal RO methodology (see pros and cons of each in sections following), the next step in design is to determine performance of available target detection reagents in presence and absence of drug. With the three basic assay designs, the operator must identify either a non-competitive clone for total receptor density, a secondary development antibody for directly bound drug detection, which can be combined with a pre-drug saturation to determine total receptor density, or a competitive clone that will bind free sites only. For modulatory/functional assays, one must investigate the ideal readout detection reagent. Ideally, the operator would find both a competitive and non-competitive clone to determine both occupancy and free sites in the same sample. This is not always an option due to binding epitopes and commercial availability of the receptor markers or due to mechanism of action resulting in internalization, shedding, etc., of the receptor. Although these are not always identified, drug manufacturers often screen candidate in-house clones for potential use in RO assays, or produce anti-idiotype antibodies that may be utilized to this end. Additional considerations must be taken if the assay for a target is to be used with a combination therapy in clinical trial. It is possible for the combinatorial therapy to affect overall clone performance. The operator must screen commercially available clones that have been conjugated to the desired fluorochrome. The ideal fluorochrome for the reagent will be dependent on expected expression levels of the target receptor. The rule is to utilize a higher staining index (bright fluorochrome) fluorochrome for lower expressing markers.

In the example shown in Table 1, a panel was designed specifically for a dual therapy model which combined an anti-PD-1 and anti-CD38 therapy. In order to perform clone screening, whole blood was taken from a normal healthy volunteer and isolated to PBMCs using a density gradient. The PBMCs were then stimulated for 24 h with PHA-L to upregulate expression of both receptors. The samples were then incubated individually or in combination with the targeted therapies for 1 h at ambient temperature. A clone of PD-1 previously determined to be an acceptable non-competitive clone was used for PD-1 detection. Two unique clones of CD38 antibodies conjugated to PE were utilized for anti-CD38 therapy competition testing. One reagent was designated as multi-clonal and designed specifically for use with anti-CD38 therapies in multiple myeloma (Cytognos, CD38 Multi-epitope FITC). The other was another commercially available clone (T16) from Beckman Coulter. All cells were co-stained with CD3 for T-cell specific gating. All reagents had previously been titrated for optimum saturating concentrations. Though the Cytognos multi-

Table 1
Antibody clone screening on PHA-stimulated PBMC

Sample matrix	Tube #: Purpose	Detection reagent	Phenotype markers	PD-1-PE (M1H4)	CD38 FITC (CYT-38F2 multi-clonal)	CD38 FITC (T16) multi-clonal)	% Competition detected based on %+ (PD-1 / CYT38/T16)
PHA Stimulated PBMCs	T1: FMO	NA	CD3				NA
	T2: Undrugged	Commercial conjugates	CD3				NA
	T3: Pre-saturate with anti PD-1 drug	Commercial conjugates	CD3				9.4% / 0.0% / 0.0% (represents avg. of 3 experiments)
	T4: Pre-saturate with anti-CD38 drug	Commercial conjugates	CD3				1.9% / 6.0% / 0.0% (represents avg. of 3 experiments)
	T5: Pre-saturate combined anti CD38 + anti PD-1 drug	Commercial conjugates	CD3				14.3% / 6.9% / 0.0% (represents avg. of 3 experiments)

epitope reagent works suitably for population gating to determine semi/quasi quantitative expression as measured by percent positive expression, it is not suitable for use in anti-CD38 RO assays with the therapy utilized in this example (Tube# T4 and T5 compared to Row label T2, column 6). The low level competition detected (6.0% and 6.9%) would result in underestimating total receptor numbers. Likewise, the PD-1 receptor detection antibody that had previously been cleared for use with this drug as a best option demonstrated interference when combined with the anti-CD38 therapy (9.4% competition anti-PD1 drug alone, 14.3% combined anti-PD1+ anti CD38). The results below were based on the average of three unique experiments in triplicate and showed how competition testing is crucial to selecting optimal reagents for monitoring target engagement without interference.

3.2 Utilizing Drug-Fluor Conjugates for Free Site Detection

Often the simplest method of detecting free sites is to utilize a fluorescent or biotinylated conjugate of the drug product. The concerns with this approach include ensuring that conjugation chemistry does not affect the binding site, does not alter the affinity, and is reproducible for lot changes. For PE and APC conjugations which have targeted conjugation chemistry, there is little lot-to-lot issue with this approach. For amine reactive

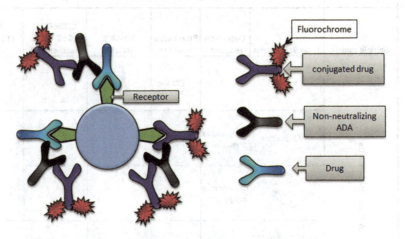

Fig. 1 Non-neutralizing ADA can result in an overestimation of free receptor by linking the detection reagent (conjugated drug) to the drug bound to the receptor or by linking together multiple detection reagents on free receptor

conjugations (i.e., Alexa Fluor™ dyes), this could mean dye in the biding site of the antibody with new lot productions [13]. This necessitates the thorough quality control of each new lot in presence and absence of drug. The direct drug-fluor conjugate method of free site detection should only be considered when either neutralizing or non-neutralizing anti-drug antibody production is unlikely. In the presence of ADA, occupancy may be inaccurately estimated, leading to under or overdosing of patients. Figures 1 and 2 demonstrate the interference caused in both instances. Thus, receptor occupancy data must be combined with sufficient PK and ADA testing to fully understand the therapeutic potency.

In addition to ADA concerns, the drug-fluor conjugate can also displace the unconjugated therapeutic over time, which can lead to an overestimation of free sites, in turn, underestimating the amount of bound drug. If these data are used for patient dosing decisions, it can lead to the investigator delivering more drug than necessary to the patient, possibly even resulting in safety and tolerability failures. In a study of MEDI3185, a monoclonal therapeutic which targets CXCR4, the assay design used AlexaFluor™ 647 conjugated MEDI3185 for free site detection and a non-competing CXCR4 commercial clone for detection of total receptors [3]. They found that after multiple ascending doses in the cynomolgus monkey model, even though PK results indicated there was no drug present in the plasma, the RO assay detected no free receptors. The corresponding testing for ADA was positive, indicating that the

Neutralizing ADA Assay Interference

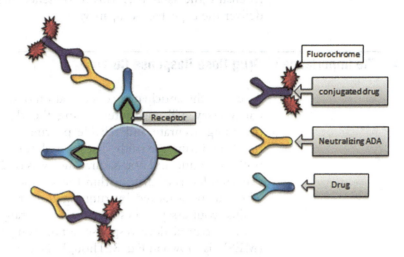

Fig. 2 In the presence of neutralizing ADA, underestimation of free receptors occurs due to detection reagent (conjugated drug) being blocked from binding to free receptor by direct interference

ADA was neutralizing the conjugated drug-fluor detection reagent [3]. These combined results exemplify the necessity to review data as a whole to understand the full therapeutic window.

3.3 Signal Amplification

As mentioned previously, for low density receptors, signal amplification is the best option for increasing the dynamic range of the assay. Utilizing a biotinylated reagent allows for more streptavidin-fluor secondary molecules to adhere to the biotinylated detection reagent than could be achieved by a direct conjugate. The amplification by biotinylation could be achieved by utilizing a biotinylated drug (free sites) or other competing (free sites) receptor detection reagent. The receptor or drug is first bound using the biotinylated reagent by incubated drugged cells with the biotinylated detection reagent, cells washed, and then a secondary streptavidin-fluorochrome conjugate is incubated with the cells and binds the biotin on the receptor detection reagent. Free secondary is washed away, and then the samples are acquired. Fluorochromes that deliver the optimum intensity for low copy number receptors include many of the Brilliant Violet™ dyes commercially available, Phycoerythrin (PE), Allophycocyanin (APC), and some Alexa-Fluor® dyes. The dye performance must be balanced with the filter configuration of the cytometer to be used for sample acquisition. Most antibody conjugate vendors have spectral calculators in which information on optimum dyes is delivered after one enters their particular instrument filter configuration into the calculator. In all cases, titrations of the detection and secondary stain reagents must be performed with suitable isotype, isoclonic and/or FMO controls to ensure the assay performs in presence and absence of drug to deliver the expected occupancy.

4 The Importance of Drug Dose Response Curves

Assessing the anticipated concentration of drug the receptor occupancy assay will encounter during the clinical trial is critical to providing accurate and reliable pharmacodynamic assessment of the drug during the study. The assay should be capable of assessing both saturating and non-saturating levels of drug and do so for the expected levels of drug within the sample matrix being collected. This can be achieved by running dose response curves of drug against your assay to best determine the range of sensitivity of the assay. A typical dose response curve using fluorescence intensity (MESF) is shown in Fig. 3. Though the dynamic range as measured by the MESF is seemingly low, these data clearly demonstrate the assay performs well over the anticipated therapeutic range expected to be achieved in patients. This is an example of a bound assay utilizing an anti-human IgG4 secondary to detect the drug. PD-L1 qualifies as a low expresser, so the signal amplification achieved with the secondary development moved the assay from one in which there was no detection window to one that can be detected within the sample matrix tested in a trial.

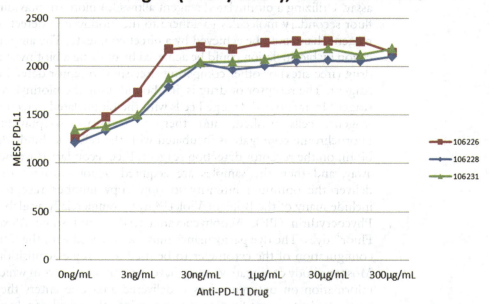

Fig. 3 Dose response curve for monitoring PD-L1 target engagement using an anti-human secondary detection of bound drug

5 Special Considerations for Assay Interference

Assay interference is any factor that inhibits the assay from performing as intended. We have already touched on some of the causes of assay interference such as ADA and conjugation chemistry, however other factors that can contribute to assay variance and even failure include issues such as:

- Impact of conjugation of drug (steric hindrance, change of binding affinity, etc.,).
- Potential dissociation of drug due to higher affinity of binding reagent or high concentrations of detection reagents.
- For functional RO assays, triggering of functional response by surface markers instead of challenge stimuli can result in inaccurate RO estimations.

6 Assay Stability Considerations

Establishing the stability of receptor occupancy assays is critical to understand the variability that may be introduced due to time between when the sample is collected and received and tested at the laboratory site. This may be less of a concern in situations when sample testing occurs in the vicinity of where patients are being evaluated, i.e., in a Phase I unit. However, when running these assays in global clinical trials with multiple clinical sites sending samples into a central testing location, you must evaluate the limits of sample stability and establish acceptance criteria around stability of your assay.

7 Blood Collection Tube Type

The choice of anti-coagulant can impact the stability of the antigen/receptor targets and should be carefully evaluated. While some assays may be compatible with the latest collection tubes containing sample preservation/stabilization reagents, care should be taken to insure they do not change the binding capacity of detection reagents to their target receptors or the actual expression profile of detected epitopes. Tube types containing low levels of formaldehyde based fixatives are not recommended for low density antigens.

8 Temperature

While it is ideal if samples for testing can be shipped at ambient temperatures, not all assays are amenable to those conditions. Stability of target receptors may be impacted by temperature once the sample is collected, especially in the case of those receptors that may be subject to some level of internalization and/or cycling once drug is administered. While shipping samples refrigerated may reduce some of the internalization observed, some receptors such as CD22 on B-cells are notorious for internalization shortly after sample collection and will require alternative means to slow or stabilize their expression during shipment. Tube type selection combined with stability testing with variable temperature conditions is crucial.

9 Semi/Quasi-Quantitative Versus Quantitative Assays

In most cases, the reportable results for flow cytometric methods are expressed as percent positive or absolute counts. These results are considered quasi-quantitative [14]. Linearity verification is not applicable to semi-quantitative/quasi-quantitative methods. Only when the fluorescence intensity signal output is quantified using fluorescence calibration/quantitation beads [15] are the results considered as quantitative and in which case linearity of the quantitation beads should be evaluated. There is risk associated with only utilizing percentages of positivity to determine occupancy or free-sites. In Fig. 4, PHA-stimulated PBMC were incubated with a saturating amount of a monoclonal antibody therapy targeting CD38 (panel b). A control set was incubated with carrier only (panel a). The cells were washed free of unbound drug, then immediately stained for CD3 and CD38 utilizing one of the clones the drug developer had reported as non-competitive based on percentages of positivity. The cells were washed a final time to remove unbound antibody conjugates, then fixed with 1% paraformaldehyde prior to acquisition on a BD FACSCanto II with a custom 3/2/3 laser configuration. As you can see from Fig. 4, when only assessing the semi/quasi quantitative readout, it appears there is no competition of the clone (95.9% positive in control cells and 95.8% positive in the drugged cells). However, if one measures the quantitative expression level based on the slope and intercept of a quantitative microsphere (bead) standard, it is apparent that the MESF of the CD38 is significantly lower in the drugged sample (MESF of 496,972 drugged vs. 743,739 undrugged), indicating there is partial competition with the clone tested. While this might be ok in targets of very high copy number, it would not be ideal for lower expression targets and would further decrease dynamic range, and lend to the underestimation of the total number of receptors. The bead standards must represent the fluorescent parameter being measured, would ideally have assigned NIST (National Institute of Standards and Technology) traceable equivalent number of reference fluorophores (ERF) or molecules of equivalent soluble fluorescence (MESF) values, and must be acquired on the same

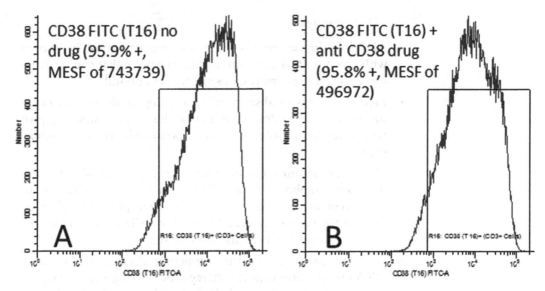

Fig. 4 Frequency and MESF of CD38 expression in drugged and undrugged samples stimulated with PHA and incubated with carrier (panel **a**) or anti-CD38 drug (panel **b**). Though percentages of this non-competitive (total target) expression are nearly identical, the MESF values are significantly different indicating importance of both types of measurements when testing for clone performance in presence of drug

instrument, at the same voltages on the same day as the samples for which the slope and intercept would be applied. If no MESF/ERF beads are available in the desired fluorochrome, then a suitable alternative would be spectrally matched standards representative of similar fluor. As long as the beads are acquired in a standardized method, this allows for accurate quantitation. Some acceptable options for quantitation beads could be, but are not limited to Bangs Laboratory MESF beads, Spherotech Ultra Rainbow Calibration Particles, and BD QuantiBrite™ beads.

10 Free Site Receptor Occupancy Assays

Advantages

- Simple development.
- Requires minimal sample.
- Pre-saturating drug incubations are not required, but can be used for background determination in absence of proper isotype.
- Similar affinity between conjugated and unconjugated drug minimizing risks of displacement.

Disadvantages

- ADA may impact results. Detection reagents must be chosen with care to MOA and combination therapies that could illicit a more robust immune response to the compound.
- Free site assays in absence of total receptor density measurements require pre-dose samples in order to determine receptor occupancy for all subsequent time points within a longitudinal study.
- Does not account for changes in receptor density beyond pre-dose sampling. If the MOA of the drug involves up or down regulation/internalization of the receptor [16], ablation of receptor-expressing cells [17], or mobilization of receptor-expressing cells from circulation to tissue.
- Free site assays require the competing clone to demonstrate 100% competition with the therapeutic binding site, and detection reagent interference balanced with assay window timing must be determined.

10.1 Free Site Receptor Occupancy: Assay Design

For free site RO assays, the background is defined as the level of fluorescence intensity and percentage of positivity when there is 100% saturation of all receptors with unconjugated drug; thus no free sites are available for a competitive antibody to bind. Free receptors are measured at each post-dose time point in absence of pre-saturation (Fig. 5).

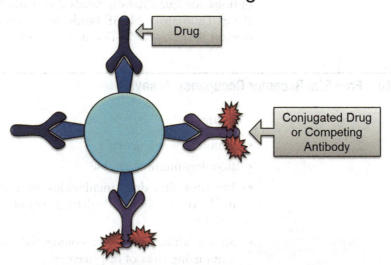

Fig. 5 In this simplified design, there are free receptors that have been detected by the competitive antibody or drug conjugate

10.2 Materials: Background Determination in Free Site ROA

1. Therapeutic drug.
2. Fluorescently conjugated competitive antibody clone or drug.
3. Fluorescently labeled non-competitive clone (optional).
4. Peripheral whole blood that expresses therapeutic target (typically sodium heparin or CytoChex collection tubes for best stability. Rarely EDTA or sodium citrate collection tubes).
5. Applicable co-stains to identify the cell population upon which the therapeutic target is expressed (i.e., CD3, CD4 and CD8 for T cell subsets, CD19 for B cell subsets).
6. Whole Blood Lysing Reagent: (i.e., ammonium chloride lysis buffer, Thermo Fisher Scientific Catalog number: 00-4333-57, or equivalent).
7. Stain Buffer: 1× DPBS with 1% BSA or FBS.
8. Fc Block, Human TruStain FcX™, BioLegend catalog number 422302, or equivalent.
9. Fixation buffer, BD Cytofix ™ Fixation Buffer, catalog number 554655, or equivalent 1–2% paraformaldehyde buffer solution.
10. Centrifuge with adaptors for 12 × 75 mm tubes, capable of 500 × g.
11. MESF/ERF quantitation standards (from vendor of choice). Should be accompanied by NIST traceable assignments if at all possible.

10.3 Method: Background Determination in Free Site ROA

1. Deliver 100 μL of anticoagulated whole blood (adjusted cell count of 10–20 × 10^6 cells/mL) to twelve 12 × 75 mm tube. For rare events, a higher starting volume may be required.
2. Add 4 mL of 37 °C pre-warmed 1× ammonium chloride RBC lysis buffer to each tube, cap, invert ten times to mix, then incubate at ambient temperature in the dark for 5 min. Centrifuge for 5 min at 500 × g at ambient temperature. Decant supernatant.
3. Add 3 mL of cold stain buffer to each tube. Tap vortex to mix. Centrifuge at 500 × g for 5 min and decant supernatant.
4. To each of six tubes (saturation tubes), add a saturating concentration of the unconjugated (cold) therapeutic to each tube, tap vortex and incubate at ambient temperature in the dark for 1 h (please note that different therapeutics may require different incubation conditions. This must be determined through prior experimentation to ensure saturation is achieved).
5. Add 3 mL of cold stain buffer, tap vortex to mix and centrifuge at 500 × g for 5 min. Decant and repeat the wash one additional time.

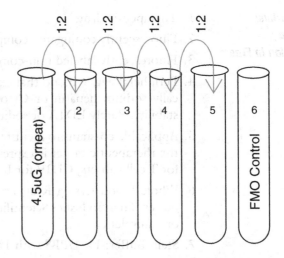

Fig. 6 Example serial dilution experiment design for background determination

6. Add applicable co-stain antibodies (i.e., CD3, CD4, CD19, etc.) in different fluorochrome than the free site antibody. Tap vortex gently to mix. Incubate at ambient temperature in the dark for 20 min.

7. Prepare a set of titration dilution tubes as indicated in Fig. 6. Begin serial titration of competitive (or conjugated drug) in the tubes as per example below. Tube 1 will contain the manufacturer's recommended staining volume of the competitive antibody. If no recommended volume is provided, start at 4.5 μg per stain reaction in Tube 1. After all dilutions, this will result in a final starting titration of 1 μg/stain reaction for the first dilution tube.

8. Add enough stain buffer to the Tube 1 such that there is 50 μL total volume in the tube (buffer volume + competitive clone volume = 50 μL).

9. Add 25 μL stain buffer to each of the other Tubes 2–5.

10. Mix the contents of Tube 1 well, and then remove 25 μL from Tube 1 and place into Tube 2. Mix well, and then remove 25 μL from Tube 2 and place into Tube 3. Repeat the process until all tubes contain two-fold serial dilutions of competitive antibody in a total of 25 μL.

11. To each of the corresponding tubes 1–6 containing the co-stained cells, add 10 μL of the corresponding competitive antibody dilution. Add 10 μL stain buffer to the FMO Tube 6.

12. Tap vortex to mix and incubate at ambient temperature (or 2–8 °C if drug requires lower temperature) in the dark for 30 min.

13. Add 3 mL of cold wash buffer and tap vortex to mix. Centrifuge at 400 × *g* for 5 min. Repeat wash one additional time.
14. Resuspend the stained cells in 500 µL fixative (2–4% paraformaldehyde).
15. Acquire 30,000 events per tube using applicable settings on a calibrated cytometer. If a rare event population is being evaluated, acquire enough events that at least 2% of the population of interest is acquired.
16. Analyze for both the percentage of positivity of the analyte of interest and Equivalent Relative Fluorescence (ERF) or Molecules of Equivalent Soluble Fluorescence (MESF). This requires acquisition of appropriate ERF or MESF quantitation beads for the channel of interest run at the same voltage as samples.
17. The titration point chosen for the competitive clone will be the point in which the MESF/ERF expression in an undrugged sample is at the plateau of the curve (i.e., at saturation) while exhibiting maximal signal-to-noise when plotting stain concentration against MESF/ERF intensity (Fig. 7). If the gating for positivity will employ use of a saturation tube to determine where to set the background, using a concentration of unlabeled drug that will fully saturate any free sites available, then you must at this point verify that in the presence of saturating amounts of drug the binding of the competitive clone is completely blocked, and there is no non-specific binding

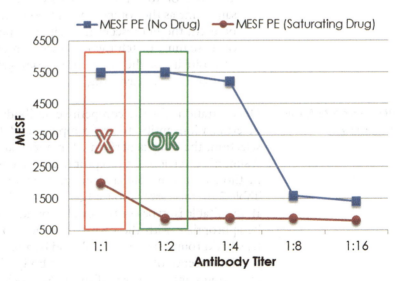

Fig. 7 Titration of the labeled competitive antibody in the presence or absence of saturating concentration of drug helps guide the appropriate titration selection for Free Site assay development. In this example, the starting titer resulted in non-specific staining that could negatively impact assessment of free sites during patient dosing

Free Site Background Determination

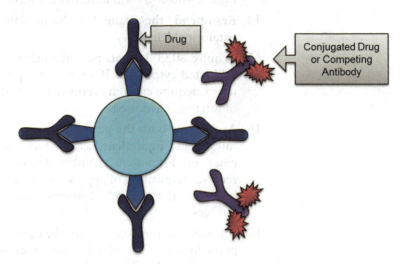

Fig. 8 The receptor targets are completely saturated ex vivo with unconjugated therapeutic. The competitive clone (which can be either the drug conjugated to a fluorochrome or another antibody that binds the exact epitope that the drug binds) is unable to bind to any receptors as they are all occupied by drug. Background = fluorescent intensity and percent positivity of any non-specifically bound fluorochrome (Red) in this scenario

present (Fig. 8). Alternatively, background can be matched to an FMO or to a non-specific isotype control titrated at the same time as the competitive antibody. If isotype or FMO is to be used, the fluorescent intensity and percentage of positivity of these must match the fluorescent intensity and percentages of positivity of the competitive conjugate in the presence of saturating unlabeled drug.

10.4 Free Site Assay Performance

Once titrations for the competitive antibody and the background determination antibody are complete and finalized for titration selection, the assay is performed in a similar manner to titrations. A simplified version of a receptor occupancy assay using the free site method measuring alpha 9 integrin on neutrophils is shown in Table 2. In this case, a mouse IgG1 was used to set the background. If no suitable isotype control can be found, then an FMO tube may be appropriate with the caveat that they do not always represent true background staining and should be assessed for accuracy versus the fluorescence of your detector antibody when in the presence of saturating concentrations of drug. An isoclonic control is also an alternative approach that can be tested when an appropriate isotype control is not identified. Failure to match this appropriately can impact calculations of RO during dosing.

Table 2
Representative data for free site RO assay using a competitive and non-competitive antibody to alpha 9 integrin

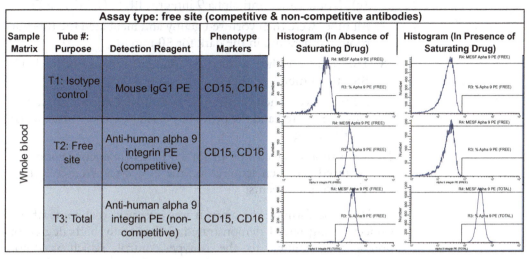

The basic layout of this panel involved three tubes as shown in Table 2. The assay evaluated alpha 9 integrin on neutrophils in peripheral blood and utilized an isotype control and custom antibodies measuring the active drug binding site and a non-competitive site on the alpha 9 integrin receptor.

The assay was processed as follows:

1. 100 μL of well-mixed Sodium Heparin anticoagulated whole blood as added to the bottom of three 12 × 75 staining tubes.

2. The red blood cells were then lysed by adding 4 mL of a whole blood lysing solution to all tubes, capping, inverting several times to mix, and incubating for 5 min at room temperature (18–22 °C).

3. All tubes were then centrifuged at 400 RCF for 5 min, supernatant decanted and the pellet dispersed by either vortexing or gentle raking to resuspend pellet.

4. The cells were then washed by adding 2 mL of PBS with 1% BSA to all tubes.

5. All tubes were then centrifuged at 400 RCF for 5 min, supernatant decanted and the pellet dispersed by either vortexing or gentle raking to resuspend pellet.

6. Pre-formulated cocktails containing the surface gating reagents were added to each tube, with the detection reagents added in the sequence below.

(a) Isotype Tube: MsIgG1 PE.
(b) Free Site Tube: Anti-alpha 9 integrin PE (FREE).
(c) Total Tube: Anti-alpha 9 integrin PE (TOTAL).

7. All tubes were then vortexed gently and incubated the tubes at room temperature in the dark for 30 min.

8. The cells were then washed by adding 2 mL of PBS with 1% BSA to all tubes.

9. All tubes were then centrifuged at 400 RCF for 5 min, supernatant decanted and the pellet dispersed by either vortexing or gentle raking.

10. Each tube received 500 μL of 1% Paraformaldehyde and was vortexed and then acquired on the flow cytometer, collecting 100,000 total events.

While performing the assay, it was observed that some of the subjects being tested demonstrated an unusually high degree of non-specific staining to the isotype control, which was not observed in normal donors used during validation. In an effort to address this observation, a mouse serum blocking step was included into the assay prior to staining with the other reagents. This immediately removed any further observation of this non-specific binding. Example data of the before and after impact is shown in Table 3.

Column A: This represents the data generated in a normal donor during validation and what is expected in the absence of drug. The percentage positive statistical region is set using the isotype control in Tube 1. Tubes 2 and 3 represent the amount of binding of the competitive and non-competitive antibodies.

Column B: This represents the data generated from a study subject prior to receiving drug where very high background is observed in Tube 1 (Isotype Control). Because of this high background, the statistical gate region is set too high, and the resulting

Table 3
Representative data of non-specific background staining and troubleshooting experiments

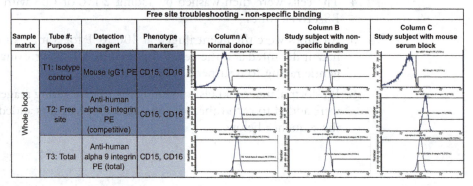

data in Tubes 2 and 3 appear to show very little positivity, even in the non-competitive antibody in Tube 3.

Column C: This represents the data from the same sample shown in *Column B*; however, this sample was pre-incubated with mouse serum. Now the background has returned to what is expected in Tube 1, and setting the percent positive stat region now permits an accurate assessment of expression expected in Tubes 2 and 3, similar to that seen in the normal subject in *Column A*.

11 Bound Site Receptor Occupancy Assays

Advantages

- Signal amplification with a secondary detection antibody or Biotin + Streptavidin secondary improves the dynamic range of the assay for low copy number receptors (Fig. 9).

Disadvantages

- Can be a two-step staining process that could introduce additional potential for error.
- Often difficult to find suitable background determination method (Fig. 9).

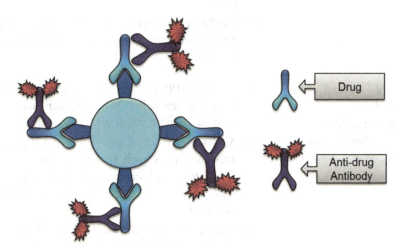

Fig. 9 The receptor targets are completely saturated ex vivo with unconjugated therapeutic. Excess cold drug is washed away and then the secondary fluorochrome-coupled detection reagent (Fc specific, Hinge region specific or an anti-idiotype) is added. The only option for background determination in this example is (1) pre-dose sample, or (2) non-specific IgG isotype

11.1 Materials: Background Determination in Bound Receptor Occupancy Assays

1. Therapeutic drug.
2. Fluorescently conjugated anti-human IgG secondary detection antibody. This will bind directly to the bound drug for detection and amplification of signal. The exact form to use will depend on the construct design of the therapeutic. The secondary can be directed against the hinge region, the Fc region, or custom anti-idiotype that only binds to bound drug. This can also be a non-competitive clone as determined by clone screening experiments.
3. Peripheral whole blood that expresses therapeutic target (typically sodium heparin or CytoChex collection tubes for best stability. Rarely EDTA or sodium citrate collection tube.
4. Applicable co-stains to identify the cell population upon which the therapeutic target is expressed (i.e., CD3, CD4 and CD8 for T cell subsets, CD19 for B cell subsets).
5. Whole Blood Lysing Reagent: (i.e., ammonium chloride lysis buffer (Thermo Fisher Scientific Catalog number: 00-4333-57, or equivalent).
6. Stain Buffer: 1× DPBS with 1% BSA or FBS.
7. Fc Block, Human TruStain FcX™, BioLegend catalog number 422302, or equivalent.
8. Fixation buffer, BD Cytofix ™ Fixation Buffer, catalog number 554655, or equivalent 1–2% paraformaldehyde buffer solution.
9. Centrifuge with adaptors for 12 × 75 mm tubes, capable of 500 × g.
10. MESF/ERF quantitation standards (from vendor of choice). Should be accompanied by NIST traceable assignments if at all possible.

11.2 Method: Background Determination in Bound Receptor Occupancy Assays

1. Deliver 100 μL of anticoagulated whole blood to twelve 12 × 75 mm tube. Note that for rare events, a higher starting volume may be required.
2. Add 4 mL of 37 °C pre-warmed 1× ammonium chloride RBC lysis buffer to each tube, cap, invert ten times to mix, then incubate at ambient temperature in the dark for 5 min. Centrifuge for 5 min at 500 × g at ambient temperature. Decant supernatant.
3. Add 3 mL of stain buffer to each tube. Tap vortex to mix. Centrifuge at 500 × g for 5 min and decant supernatant.
4. To each of six tubes (saturation tubes), add a saturating concentration of the unconjugated (cold) therapeutic, tap vortex and incubate at ambient temperature in the dark for 1 h (please

note that different therapeutics may require different incubation conditions. This must be determined through prior experimentation to ensure saturation is achieved).

5. Add 3 mL of stain buffer, tap vortex to mix and centrifuge at 500 × g for 5 min. Decant and repeat the wash one additional time.

6. Add applicable co-stain antibodies (i.e., CD3, CD4, CD19, etc.) in different fluorochrome than the bound site antibody. Tap vortex gently to mix. Incubate at ambient temperature in the dark for 20 min.

7. Prepare a set of titration dilution tubes as indicated below. Begin serial titration of secondary detection reagent in the tubes as per example below. Tube 1 will contain the manufacturer's recommended staining volume of the secondary detection antibody. If no recommended volume is provided, start at 4.5 µg per stain reaction in Tube 1. After all dilutions, this will result in a final starting titration of 1 µg/stain reaction for the first dilution tube.

8. Add enough stain buffer to the Tube 1 such that there is 50 µL total volume in the tube (buffer volume + competitive clone volume = 50 µL).

9. Add 25 µL stain buffer to each of the other Tubes 2–5.

10. Mix the contents of Tube 1 well, and then remove 25 µL from Tube 1 and place into Tube 2. Mix well, and then remove 25 µL from Tube 2 and place into Tube 3. Repeat the process until all tubes contain two-fold serial dilutions of secondary detection antibody in a total of 25 µL.

11. To each of the corresponding tubes containing the co-stained cells, add 10 µL of the corresponding secondary detection antibody dilution. Tap vortex to mix and incubate at ambient temperature (or 2–8 °C if drug requires) in the dark for 30 min.

12. Add 10 µL stain buffer to the FMO tube.

13. Add 3 mL of cold wash buffer and tap vortex to mix. Centrifuge at 400 × g for 5 min. Repeat wash one additional time.

14. Resuspend the stained cells in 500 µL fixative (2–4% paraformaldehyde).

15. Acquire 30,000 events per tube using applicable settings on a calibrated cytometer. If a rare event population is being evaluated, acquire enough events that at least 2% of the population of interest is acquired.

16. Analyze for both the percentage of positivity of the analyte of interest and Equivalent Relative Fluorescence (ERF) or Molecules of Equivalent Soluble Fluorescence (MESF). This

requires ERF or MESF intensity particles channel of interest, run at same voltage as samples.

17. The titration point chosen for the secondary detection reagent will be the point in the saturation drugged that demonstrates the highest signal to noise ratio (if applicable), with a plateau of the MESF/ERF. Background can be matched to an FMO or to a non-specific isotype control titrated at same time as the secondary detection antibody (Fig. 10, Table 4).

12 Assay Design: Functional Receptor Occupancy Assays

This is a special format of RO testing since you are not only measuring occupied or free receptors but are measuring the dynamic change in protein expression levels in direct proportion to the occupancy of drug. Typically, the dosed samples are split with one aliquot being challenged with the stimulus that will illicit the dynamic change and the other serving as a control with no challenge. These assays can be a direct measure of receptors that are bound and still present on a cell after a receptor internalization challenge, cytokine production after a stimulus challenge to the target receptor, or a phosphorylation readout that is proportional to the occupied receptors. This design method requires a firm understanding of the mechanism of action of the drug, the function of the receptor, and any downstream events that happen after receptor engagement. There are many experimental approaches for this type of occupancy assay. Below, we cover an internalization method, but others have used in clinical trials cytokine production as a functional readout of receptor occupancy.

Advantages

- This method allows a direct functional and correlative readout of drug function in-lieu of direct occupancy. The function is often associated with the mechanism of action and can be more sensitive than traditional target occupancy.
- This method may best be used for small molecule inhibitors that covalently bind receptor or for newer small antibody fragment therapeutics such as diabodies that are not amenable to traditional detection methods.

Disadvantages

- These methods often require fixation and permeabilization that may result in altered ability to detect surface phenotypic markers.
- Often requires long incubations in sterile conditions.

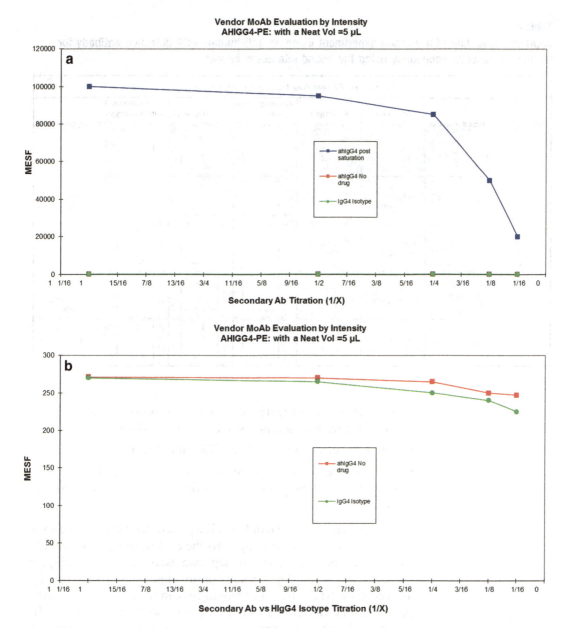

Fig. 10 Figure representing titration experiments for the secondary Ab detection reagent for monitoring bound receptor (a) against the human IgG4 isotype control (b). In this example, the isotype and anti-human IgG4 secondary detection antibody for an occupied receptor assay has been titrated on matched donors in presence (blue line) and absence of drug (red and green lines). The correct titration point for the secondary is chosen first based on the MESF on the pre-saturated sample (a). The titration point for the isotype is the point at which the MESF of the isotype (Green line) most closely matches the anti-human IgG4 (red line) in absence of drug (b). In the examples above, the anti-human IgG4 titration is called at 1:2, and the IgG4 isotype is best utilized at 1:1

Table 4
Representative data of a titration experiment using an anti-human IgG4 detection antibody for monitoring receptor occupancy using the bound site assay format

Sample Matrix	Tube #: Purpose	Detection reagent	Phenotype markers	A. Histogram (in presence of saturating drug)	B. Histogram (in absence of drug)	C. Histogram (matched isotype control)	Calculation option
Whole blood	T1: Neat	Anti-human IgG4 fluor conjugate)	CD3, CD4, CD8, CD19)				
	T2: 1:2 of neat	Anti-human IgG4 fluor conjugate)	CD3, CD4, CD8, CD19)				%Occupied Receptor = [(B-C)/(A-C)]*100
	T3: 1:4 of neat	Anti-human IgG4 fluor conjugate)	CD3, CD4, CD8, CD19)				
	T4: 1:8 of neat	Anti-human IgG4 fluor conjugate)	CD3, CD4, CD8, CD19)				A = MESF αhIgG4 total receptor
	T5: 1:16 of neat	Anti-human IgG4 fluor conjugate)	CD3, CD4, CD8, CD19)				B = MESF αhIgG4 occupied receptor C = MESF hIgG4 Iso

- Dose response of the stimulator is critical as overstimulation may mask the measured functional effect of the drug.
- Often requires CO_2 and humidified incubators.
- Stability may be of concern and must be stringently tested to determine at which time point ex vivo that results are no longer valid.

In the cartoon shown in Fig. 11, a patient has been treated with the therapy, then sample shipped to the testing facility. The sample would then be aliquoted into separate assay tubes. One aliquot would receive saturating amounts of unconjugated drug to occupy all receptors, while the other would remain undrugged. The samples would be challenged with a stimulus that elicits a response. In the example below, the response is that all receptors that are free of drug would be internalized. An anti-drug antibody would then be used to determine the expression levels of receptor on the surface of the cell that did not internalize. In this example, the saturation tube is equivalent to total receptor expression, as the receptors cannot internalize when bound by drug. The control tube would be the measure of bound expression levels. Combined results from the two tubes can be used to determine % RO.

Precision Medicine: The Function of Receptor Occupancy in Drug Development 193

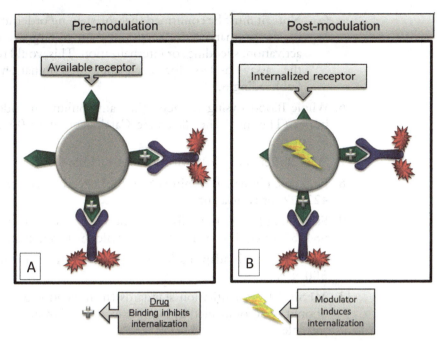

Fig. 11 Example of possible challenge/internalization assay that can be used to determine receptor occupancy

12.1 Materials: Background Determination in Modulatory/Functional Internalization ROA

1. Therapeutic drug.
2. Challenge reagent that induces receptor internalization.
3. Fluorescently conjugated anti-human IgG secondary detection antibody. This will bind directly to the bound drug for detection and amplification of signal. The exact form to use will depend on the construct design of the therapeutic. The secondary can be directed against the hinge region, the Fc region, or custom anti-idiotype that only binds to bound drug. This could also be a non-competitive clone as determined by clone screening experimentation.
4. Peripheral whole blood that expresses therapeutic target (typically sodium heparin, sodium citrate, or EDTA.
   (a) Note that collection tubes containing a fixative to stabilize expression cannot be utilized for a functional assay of this sort. The stabilizers in the commercially available tubes, such as CytoChex and others contain a small percentage of paraformaldehyde which crosslinks the surface markers. This renders the cells non-viable and unable to respond to stimuli.
5. Applicable co-stains to identify the cell population upon which the therapeutic target is expressed (i.e., CD3, CD4 and CD8 for T cell subsets, CD19 for B cell subsets).

(a) Note: it must be confirmed that the staining order utilized for surface markers does not contribute to target receptor activation, shedding, or internalization. This would render the assay useless for occupancy determination. *See* Subheading 5.

6. Whole Blood Lysing Reagent: (i.e., ammonium chloride lysis buffer (Thermo Fisher Scientific Catalog number: 00-4333-57, or equivalent).
7. Stain Buffer: 1× DPBS with 1% BSA or FBS.
8. Fc Block, Human TruStain FcX™, BioLegend catalog number 422302, or equivalent.
9. Fixation buffer, BD Cytofix ™ Fixation Buffer, catalog number 554655, or equivalent 1–2% paraformaldehyde buffer solution.
10. Centrifuge with adaptors for 12 × 75 mm tubes, capable of 500 × *g*.
11. MESF/ERF quantitation standards (from vendor of choice). Should be accompanied by NIST traceable assignments if at all possible.

12.2 Methods: Background Determination in Modulatory/Functional Internalization ROA

1. Split an aliquot of whole blood into two 15-mL conical tubes. Add drug in saturating concentrations to one aliquot, and add carrier control to the second aliquot. Place both in 37 °C water bath for 1 h.
2. Deliver 100 μL of anticoagulated, well mixed whole blood to six 12 × 75 mm tube. Note that for rare events, a higher starting volume may be required. Ensure that the saturated condition tubes are labeled adequately.
3. Add challenge reagent to each tube in the predetermined concentration for the predetermined incubation period. This will be determined via kinetic response experimentation as a pre-validation step.
4. Add serially diluted receptor detection reagent that has been prepared as previously described in other ROA sections within this chapter. Add any other surface phenotypic antibodies required for gating at this point. Incubate in the dark at refrigerated condition for 20 min.
5. Add 4 mL of 37 °C pre-warmed 1× ammonium chloride RBC lysis buffer to each tube, cap, invert ten times to mix, then incubate at ambient temperature in the dark for 5 min. Centrifuge for 5 min at 500 × *g* at ambient temperature. Decant supernatant. Alternatively, use a combine lysis/fixation buffer such as BD Lyse/Fix if shedding of detection reagent or drug is suspected. This will help stabilize both.

6. Add 3 mL of stain/wash buffer to each tube. Tap vortex to mix. Centrifuge at 500 × g for 5 min and decant supernatant. Repeat one additional time.

7. Resuspend stained cells in 1–2% paraformaldehyde containing fixation buffer and acquire on cytometer of choice immediately.

8. Analyze each titration point by gating for positivity first on the undrugged, challenged titration tubes and applying that to the matching titration point tubes in the challenged drugged and undrugged samples. The isotype control should then be matched to the same undrugged, challenged tubes to pick the optimum titration point to be used.

12.3 CCR5 Specific Internalization RO Assays

The method described above can be utilized for any drug product that inhibits receptor internalization if the internalization trigger is known and available for use in the assay. CCR5 is a perfect example of a target that is amenable to this type of assay. Specific therapeutics, including pertussis toxin, are able to inhibit the internalization of the receptor if bound to it. A challenge with MIP-1α induces rapid internalization of the unbound receptors. Occupancy of drug can then be determined by incubating the cells with the non-competitive clone of the anti-receptor antibody that is fluorochrome conjugated [18]. The resulting percentage of CCR5+ cells is a direct measure of target occupancy.

In the example shown in Fig. 12, a single level dose of therapeutic to CCR5 was delivered to patients followed by 19 days of whole blood collection. The blood was tested within the challenge method described above and assessed for %CCR5 on CD4+ T cells. The occupancy increased rapidly after administration and remained stable until Day 15, then began to decline. This assay combined with PK data helped the investigator determine the time of next cohort dosing.

13 Conclusion

Research and clinical development of biologics aimed at modulating the immune response increased significantly over the years. Drugs of this class have been developed to deliver cytotoxic molecules, redirect the cellular arms of the immune response and/or directly modulate the host's immune response by enhancing or diminishing immune effector functions. With the increased approval of immune-modulating therapies, we are witnessing many successful treatments against Cancer, Autoimmune, and Infectious diseases. Flow Cytometry has been widely used for monitoring the function and phenotype of cells with rapidity and precision. To gain a better understanding of the therapeutic activity of biologics, customized cellular assays such as receptor occupancy

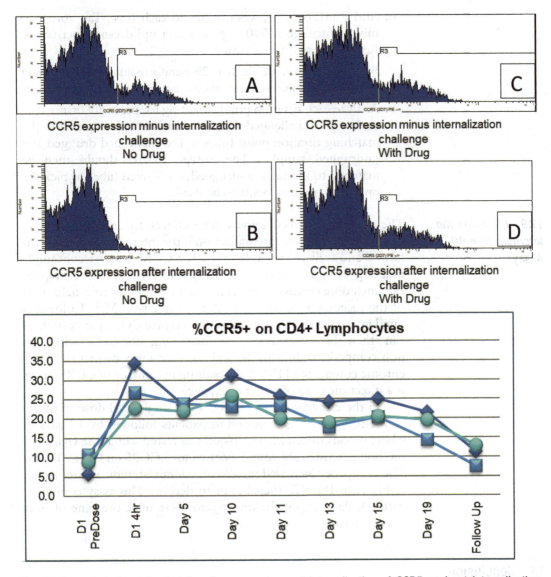

Fig. 12 Representative data depicting the expression and internalization of CCR5 pre/post internalization challenge with MIP-1α in the presence/absence of drug

have been validated and applied at all phases of the drug development cycle. Receptor occupancy assays, when properly designed, validated and operationalized under controlled conditions; serve as a powerful tool in the assessment of therapeutic binding and activity. In this Chapter, we aimed to describe the various assay formats and point to the critical parameters to take into consideration during development and validation of receptor occupancy.

References

1. Green CL, Stewart JJ, Höerkorp C, Lackey A, Jones N, Liang M, Xu Y, Ferbes J, Moulard M, Czechowska K, McCloskey TW, van der Strate BWA, Wilkins DEC, Lanham D, Wyant T, Litwin V (2016) Recommendations for the development and validation of flow cytometry-based receptor occupancy assay. Cytometry B 90B:141–149
2. Stewart JJ, Green CL, Jones N, Jones N, Liang M, Xu Y, Wilkins DEC, Moulard M, Czechowska K, Lanham D, McCloskey T, Ferbas J, van der Strate BWA, Höerkorp C, Wyant T, Lackey A, Litwin V (2016) Role of receptor occupancy assays by flow cytometry in drug development. Cytometry B 90B:110–116
3. Liang M, Schwickart M, Schneider AK, Vainshtein I, Del Nagro C, Standifer N, Roskos LK (2016) Receptor occupancy assessment by flow cytometry as a pharmacodynamic biomarker in biopharmaceutical development. Cytometry B 90B:117–127
4. Suntharalingam G, Perry MR, Ward S, Brett SJ, Castello-Cortes A, Brunner MD, Panoskaltis N (2006) Cytokine strom in a phase 1 trial of the anti-CD28 monoclonal antibody TGN1412. N Engl J Med 355:1018–1028
5. Muller PY, Brennan FR (2009) Safety assessment and dose selection for first-in-human clinical trials with immunomodulatory monoclonal antibodies. Clin Pharmacol Ther 85:247–258
6. Moulard M, Ozoux ML (2016) How validated receptor occupancy flow cytometry assays can impact decisions and support drug development. Cytometry B 90B:150–158
7. Sternebring O, Alifrangis L, Christensen TF, Ji H, Hegelund AC, Höerkorp C (2016) A weighted method for estimation of receptor occupancy for pharmacodynamic measurements in drug development. Cytometry B 90B:220–229
8. Fisher TL, Seils J, Reilley C, Litwin V, Green L, Salkowitz-Bokal J, Walsh R, Harville S, Leonard JE, Smith E, Zauderer M (2016) Saturation monitoring of VX15/2503, a novel semaphorin 4D-specific antibody, in clinical trials. Cytometry B 90B:199–208
9. EMA (2007) Guideline on strategies to identify and mitigate risks for first-in-human clinical trials with investigational medicinal products. EMA, Amsterdam. http://www.ema.europa.eu/docs/en_GB/document_library/Scientific_guideline/2009/09/WC500002988.pdf
10. FDA (2005) Guidance for industry: estimating the maximum safe starting dose in initial clinical trials for therapeutics in adult healthy volunteers. FDA, Silver Spring, MD. http://www.fda.gov/downloads/Drugs/Guidances/UCM078932.pdf
11. Maecker HT, McCoy JP, Nussenblatt R (2012) Standardizing immunophenotyping for the Human Immunology Project. Nat Rev Immunol 12(3):191–200
12. Fan G et al (21 Dec. 2015) Bispecific antibodies and their applications. J Hematol Oncol 8:130
13. Szabó Á, Szendi-Szatmári T, Ujlaky-Nagy L et al (2018) The effect of fluorophore conjugation on antibody affinity and the photophysical properties of dyes. Biophys J 114(3):688–700
14. Lee JW, Weiner RS, Sailstad JM, Bowsher RR, Knuth DW, O'Brien PJ, Fourcroy JL, Dixit R, Pandite L, Pietrusko RG, Soares HD, Quarmby V, Vesterqvist OL, Potter DM, Witliff JL, Fritche HA, O'Leary T, Perlee L, Kadam S, Wagner JA (2005) Method validation and measurement of biomarkers in nonclinical and clinical samples in drug development: a conference report. Pharm Res 22(4):499–511
15. Wang L, Gaigalas AK, Marti G, Abbasi F, Hoffman RA (2008) Toward quantitative fluorescence measurements with multicolor flow cytometry. Cytometry A 73(4):279–288
16. Lacy MQ et al (2008) Phase I, pharmacokinetic and pharmacodynamic study of the anti-insulinlike growth factor type 1 Receptor monoclonal antibody CP-751,871 in patients with multiple myeloma. J Clin Oncol 26:3196–3203
17. Colombat P et al (2001) Rituximab (anti-CD20 monoclonal antibody) as single first-line therapy for patients with follicular lymphoma with a low tumor burden: clinical and molecular evaluation. Blood 97:101–106
18. Mueller A, Mahmoud N, Goedecke C, al e (2002) Pharmacological characterization of the chemokine receptor, CCR5. Brit J Pharmacol 135:1033–1043

Chapter 12

In Vitro Assays for Assessing Potential Adverse Effects of Cancer Immunotherapeutics

Jinze Li, Mayur S. Mitra, and Gautham K. Rao

Abstract

The field of cancer immunotherapy (CIT) covers a wide and ever-growing variety of molecular platforms and modalities. Since the overall aim of CIT is to activate the immune system and elicit anticancer immune responses, a majority of adverse effects noted with CIT drugs are related to the molecule's pharmacology and exaggeration of pharmacological responses. A major challenge for CIT is that nonclinical toxicity studies utilizing healthy animals often do not identify relevant pharmacological toxicities due to low or no expression of target in nontumor tissues. Therefore, the design of a battery of in vitro assays is crucial for the assessment of potential risks to cancer patients and the set-up of an adequate safety monitoring plan for clinical trials. The effective translation of results from in vitro assays to in vivo safety assessment requires a thorough understanding of the molecule's biology and mechanism of action (MoA) and careful interpretation of desired versus exaggerated pharmacology. Given the vast array of platforms and modalities for CIT, in vitro assays to assess potential adverse effects must be tailored on an individual molecule basis. In this chapter, we highlight some of the principal types of in vitro assays conducted during the course of CIT drug development, the considerations and challenges therein, and how these assays contribute to the overall safety assessment of CIT drug candidates.

Key words In vitro assays, Cancer immunotherapy, Biotherapeutics, Bispecific modalities, Drug development, Safety assessment

1 Introduction

Over the last decade, cancer immunotherapy (CIT) has come to the forefront of medicine. Although the idea of using the immune system to attack cancer is not new [1], a better understanding of both the fields of immunology and cancer biology has revolutionized the landscape of cancer treatment. Multiple clinical trials have shown that cancer immunotherapies increase survival rates of patients with cancers that typically have high mortality rates. However, to date only a relatively small percent of cancer patients respond well to this class of drugs [2]. Nevertheless, the clinical benefit shown by CIT drugs and the promise of extending that benefit to a larger number of patients via more efficacious molecules and/or more rational combination therapies have resulted in an upsurge in the number of CIT molecules in nonclinical and clinical development.

At present, CIT drug candidates cover a wide array of platforms and modalities ranging from orally bioavailable small molecules to parenterally administered engineered biotherapeutics and personalized cancer vaccines. The rapid progress in the generation of new modalities has come with its own set of challenges in the demonstration of efficacy and safety of these molecules. There is also a growing appreciation that derisking and safety assessment approaches for CIT molecules must be tailored to fit the unique biology and mechanism of action (MoA) of the molecule being developed. Since the overall aim of CIT is to activate the immune system and elicit anticancer immune responses, a majority of adverse effects noted with CIT drugs are related to the molecule's pharmacology and exaggeration of pharmacological responses. Consequently, in vitro assays used to assess potential adversities are often the same that are used to demonstrate the pharmacology and MoA during the course of development. One of the most challenging aspects of CIT drug development remains how to extract information on desired versus exaggerated pharmacology from in vitro assays and how to effectively translate in vitro findings to in vivo safety assessment.

The types of in vitro assays considered as part of safety assessment of CIT molecules should be guided by the molecule's known biology, MoA, platform, and potential safety liabilities resulting from an exaggeration of the pharmacological activity (Table 1). Potential safety liabilities for CIT drugs may include direct activation of immune cells, cytokine release, enhanced risk of infusion reactions, infection risk, and autoimmunity. In the clinic, a broad spectrum of immune-related adverse events has been noted—in some cases severe—in patients treated with CIT drugs [3]. In order to identify potential risks to patients and set up an adequate safety monitoring plan, it is crucial to be able to identify safety liabilities in the preclinical space. A unique challenge for many CIT molecules, however, is that many of the targets are expressed at low levels, or not at all, in nontumor tissues. Consequently, nonclinical toxicity studies utilizing healthy nontumor bearing animals may not identify hazards relevant to cancer patients, and risk assessment may rely instead on in vitro assays. Furthermore, in vitro activity assays may also guide the selection of the first-in-human (FIH) dose for CIT drug candidates.

The sections below summarize some of the principal types of in vitro assays conducted, the considerations in the conduct of these assessments, and how these assays may contribute to the overall safety assessment of CIT drug candidates. While we provide a brief discussion on methodologies, this chapter is not intended to provide detailed descriptions on assay protocols.

Table 1
CIT modalities and type of in vitro assays to consider during the course of nonclinical development

CIT platforms	In vitro binding/ potency	Fc effector function	Target cell killing/ effector[a] activation	Cytokine release assays[b]	TCR assay	ADA[c] assessments	Antigen[d] or viral challenge	Other immune[e] endpoints
mAb[f] or mAb fragments	Yes[g]	Yes[h]	No	Yes[i]	Yes[j]	Yes	Yes[k]	No
Bispecific modalities	Yes[g,h]	Yes[h]	Yes	Yes[i,l]	Yes[j]	Yes	No[k]	No
Small molecules	No	No	No	No	No	No	Yes[k]	No
Oncolytic viruses	No	No	Yes[m]	No[n,o]	No	No	No	Yes[p]
ADCs	Yes[g]	Yes[h,q]	Yes	Yes[i,q]	Yes[j,q]	Yes	No	No
Cancer vaccines[r]	No	No	No	No	No	No	No	Yes[s]
Cytokines	Yes[t]	No	Yes[u]	No	No	Yes	No[k]	Yes[u]
CAR[v] therapies	No	No	Yes[w]	No[w]	No	No	No	Yes[x]

ADA antidrug antibody, *ADC* antibody-drug conjugate, *CAR* chimeric antigen receptor, *TCR* tissue cross-reactivity

[a]Non-Fc-mediated immune-effector function related assays
[b]Cytokine release in soluble and/or solid phase formats in human whole blood or PBMCs
[c]ADA assessment could be considered in in vivo nonclinical studies to explain changes in exposure to drug and/or unexpected toxicities
[d]In vitro or ex vivo immune function assessments following challenges with immunogens such as KLH or viruses in in vivo nonclinical studies
[e]Immune function or immunotoxicology endpoints are conducted following a tiered approach and may be relevant to multiple modalities depending on MoA
[f]Includes mAbs directed against all targets
[g]Binding assessment to both target antigen(s) and Fc receptors
[h]If construct contains an Fc domain
[i]If target is expressed on surface of immune cells
[j]TCR assay may not be necessary if S9 indications only
[k]Considered on a case-by-case basis and may be used to demonstrate immunomodulation in the absence of other markers of activity in nonclinical toxicity studies
[l]Cytokine release for bispecific modalities may be assessed within the target cell killing/effector cell activation assays instead
[m]In vitro assay to assess lysis of target tumor cells
[n]In vitro assessment of cytokine release can be considered because of lytic MoA using alternate formats
[o]Measurement of expressed transgene (e.g., hGM-CSF) may be considered in in vivo studies as a biomarker of activity
[p]Measurement of T-cell activation by IFN-γ ELISpot or other methodology within in vivo (pharmacology/proof-of-concept) studies
[q]May not be needed if carrier antibody previously characterized
[r]Including personalized neoantigen vaccines
[s]Detection of antigen-specific T-cells by IFN-γ ELISpot or MHC tetramer assays in in vivo studies
[t]Binding to cognate receptor(s)
[u]Activation/inhibition of target cells and associated functional immunotoxicology endpoints
[v]Refers to adoptive cell therapies with transduced antigen receptors
[w]Activation of transduced cells and potential for cytokine release may be assessed in in vivo studies
[x]Assessment of functional immunotoxicology endpoints following activation of transduced cells

2 Binding Affinity and Potency Assays

When considering modalities such as monoclonal antibodies (mAbs) or multitargeting biotherapeutics, the binding affinity of the Fab portion of therapeutic monoclonal antibodies (mAbs) to the target antigen in human and nonclinical toxicology species is determined during the early discovery phases to select a viable lead clinical candidate and also to choose an appropriate toxicology species for nonclinical development. The kinetics of this interaction is determined by calculating the dissociation constant (K_D) and on-off rates (K_{on}/K_{off}). When the total concentrations of ligand and target are higher than the K_D value, most binding partners exist in the associated form. The K_D can be determined using different assays. These include enzyme-linked immunosorbent assay (ELISA), surface plasmon resonance (SPR or BIAcore), and kinetic exclusion assays (KinExA), the first two being most popular [4].

The SPR method is based on the principle of changes in refractive index when a mAb and the target antigen interact within about 150 nM from the surface of the sensor. This interaction is studied by attaching one of the molecules to the sensor surface and by continuously passing a solution of the second molecule over the surface. A special optical biosensor is applied to measure the change in refractive index when the two molecules are in the bound state. The SPR response is directly proportional to the change in mass concentration close to the surface. Labeling of the target antigen or the therapeutic mAb is not required in the SPR assays and, therefore, this assay is widely used as a high-throughput method to determine binding affinity between two molecules. However, the SPR assay has a narrow dynamic range, compared to ELISA assays, and is comparatively less sensitive in measuring interactions with affinity (K_D) values below 100 pM. In such cases, ELISA can be used to measure binding affinity (K_D) *as low as* 10 pM [4, 5]. Another assay, KinExA, has been recently developed to measure binding affinity of biomolecules. In this assay, a therapeutic mAb in solution form is added to a small column of beads, which contains an immobilized binding partner (target antigen). The free concentration of the antibody remaining in solution after having reached equilibrium is determined using a fluorescence detector. The binding affinity at equilibrium (K_D) is then calculated on the basis of a decrease in concentration of the antibody in solution [6].

The Fab portion of an antibody is mainly associated with binding to the target whereas the Fc portion may interact with FcR (Fc receptors) on immune cells and lead to complement activation and other effector functions. These effector functions include antibody-dependent cellular cytotoxicity (ADCC), antibody-dependent cell-mediated phagocytosis (ADCP), and complement-dependent cytotoxicity (CDC). Refer to Section on *Fc Effector Function Assays* for further reading.

Binding to the target and/or FcRs generally is the first step in the functioning of a therapeutic mAb. However, the MoA of an mAb is not only dependent on the binding potential but also on how effectively it modulates the downstream pharmacodynamics events. Therefore, in addition to measuring the binding affinity of a biotherapeutic, a thorough characterization of its in vitro cellular potency is critical. Importantly, in vitro cellular potency is a more relevant indicator of biological activity compared to the binding affinity [7].

The cellular potency of a therapeutic protein is measured by developing a scientifically qualified in vitro cellular assay based on the knowledge of the therapeutics target, MoA, and downstream signaling events. In these assays, the potency of the biologic is compared to a positive and a negative control [4]. The cellular potency of a biologic is defined in terms of EC_{50}, which represents the concentration at which the molecule modulates the pharmacodynamics response by 50% between baseline to maximum effect. Data from in vitro cellular potency assays are often used in modeling and simulation efforts to project clinical efficacious exposures [7]. These characteristics are evaluated during the early discovery stages and form the basis for lead candidate selection. Additionally, these data are routinely used to characterize batch-to-batch variability of manufacturing lots [8].

3 Fc Effector Function Assays

While the Fab domain of mAbs plays an important role in selectively engaging the target by blocking or mimicking receptor-ligand interactions, the Fc domain can also play a critical role in therapeutic efficacy. The Fc domain mediates effector function by interacting with Fc receptors expressed on immune cells. Interactions with Fc receptors has the potential to mediate toxicity via cytokine release or tumor lysis syndrome [4]. A thorough assessment of Fc effector function is critical early in the development of mAbs and related CIT biotherapeutics containing Fc domains. While the binding potency and affinity to a variety of Fc receptors can be assessed in cell-free systems via SPR or other methods [9], these assays do not provide information on the functional relevance of the aforementioned binding interactions. Therefore, a battery of in vitro cell-based assays is conducted to assess Fc effector function.

There are three main types of Fc effector function assays that are considered in the course of development of mAbs and related Fc-containing molecules: (1) antibody-dependent cellular cytotoxicity (ADCC), (2) antibody-dependent cellular phagocytosis (ADCP), and (3) complement-dependent cytotoxicity (CDC).

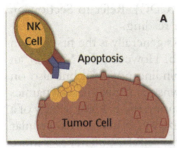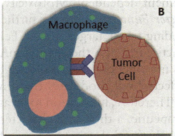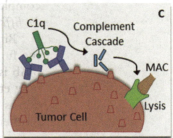

Fig. 1 Simplified schematic of immune effector functions mediated by Fc domains of mAbs: ADCC (**a**), ADCP (**b**), and CDC (**c**). Many mAbs in the CIT space are engineered to introduce modifications in the Fc domain to enhance or dampen immune effector functions depending on the intended MoA

These assays examine different immune effector functions mediated via the Fc domain (Fig. 1). In general, in vitro characterization of Fc effector function is performed in early stages of the development. Key considerations for the conduct of these in vitro assays include the isotype of antibody, the types of modifications introduced within the Fc domain, and whether Fc receptor interactions are an intended or unintended effect. Human IgG1, for example, is known to be more efficacious at eliciting Fc effector functions while other isotypes are less efficacious or not efficacious at all [10–12]. Over the last two decades, multiple antibody engineering methods have been described to enhance or silence Fc effector functionality [13]. Many mAbs in the CIT space are engineered to introduce these modifications in the Fc domain to enhance immune-mediated cytolytic effects on tumor cells or alternatively to silence this effect depending on the intended MoA. For example, afucosylation of the Fc domain of IgG1 has been shown to enhance affinity to FcγRIIIa and is part of the MoA of various marketed mAbs including obinutuzumab and mogamulizumab [14, 15]. Afucosylated versions of commonly used depleting therapeutic mAbs have superior efficacy in vivo [16, 17] and have been clinically validated. Amino acid modifications may also be introduced within the Fc domain to silence Fc effector function and prevent target cell death (on-target toxicity) or reduce off-target cytotoxicity [15, 18]. Consequently, in vitro cell-based Fc effector function assays are conducted either to demonstrate the intended Fc receptor-dependent pharmacology or to confirm a lack of unwanted Fc effector function effects. In addition, these in vitro assays also allow for an assessment of potential immune cell activation and cytokine release liabilities as a result of Fc effector function.

3.1 ADCC Assays

ADCC is an immune mechanism whereby Fc receptor-bearing immune effector cells actively engage and kill antibody-coated target cells. ADCC is a critical mechanism underlying the clinical efficacy of therapeutic anticancer mAbs and has been implicated in the MoA of rituximab, trastuzamab, alemtuzumab, and cetuximab [19]. The ADCC process involves killing target cells through a nonphagocytic process and is characterized primarily by the release of cytotoxic granules (perforin and granzymes) or secondarily by expression of apoptosis-inducing molecules such as Fas ligand on the surface of immune effector cells [20]. The ADCC activity of mAbs is mostly attributed to NK cells via the interaction Fc domain with FcγRIIIa [21, 22]. However, monocytes, macrophages, neutrophils, eosinophils, and dendritic cells are also capable of mediating an ADCC [20]. Human IgG1 and IgG3 are considered more efficacious at eliciting ADCC while human IgG2 and IgG4 have limited ADCC potential [10, 11]. ADCC assays should be considered for all CIT biotherapeutics containing an Fc domain including standard mAbs, Fc-engineered mAbs, ADCs, bispecific modalities, and other related antibody constructs.

In vitro ADCC assays normally employ human PBMCs, which naturally contain NK cells, or NK cell lines that have been activated to express high levels of FcγRIIIa as immune effector cells for IgG-Fc binding. Target cells in ADCC assays may be human PBMCs or cell lines expressing the target antigen. A traditional method for assessing ADCC activity is incubation of ^{51}Cr-loaded target cells with the therapeutic mAb followed by incubation with immune effector cells and measurement of the release of ^{51}Cr from dead target cells by scintillation counting [23]. Alternatively, target cell killing can also be assessed using nonradioactive methods such as release of the fluorescent calcein, measurement of cytoplasmic lactate dehydrogenase, or direct quantification of dead cells using carboxyfluorescein succinimidyl ester (CFSE) staining [24]. Additionally, since the Fc receptor interactions can also result in effector cell activation, NK cells may be assessed for upregulation of activation markers such as CD69 and CD107a using flow cytometry or release of perforin and cytokines into the cell medium [25]. An important factor in the design and conduct of ADCC assays is the ratio of target to effector cells. Since the receptor densities of FcγRIIIa will also likely differ from donor to donor (if using primary cells), optimization of the ratio of target to effector cells is an important component of ADCC assay development [24].

3.2 ADCP Assays

ADCP is an immune mechanism by which antibody-opsonized target cells activate Fc receptors on the surface of phagocytes resulting in the internalization and degradation of the target cell through acidification of the phagosome. ADCP can also be mediated by several other phagocytic cell types including macrophages, monocytes, neutrophils, and dendritic cells [26]. However, macrophages are the primary effector cell type implicated in ADCP. Macrophages are extremely efficient in killing circulating tumor cells via phagocytosis in the presence of tumor-specific mAbs.

Since the microenvironment in multiple solid tumors is abundant in macrophages [26], tumor cell-depleting mAbs with enhanced ADCP potential may be especially efficacious. ADCP has been implicated as a major MoA of several CIT biotherapeutics including elotuzumab, which has been shown to induce activation of macrophages in vivo and mediate ADCP of myeloma cells in an Fc receptor-dependent manner [27, 28]. In general, mAbs with enhanced binding to activating Fcγ receptors—particularly FcγRIIa, but also FcγRI and FcγRIIIa—and decreased binding to the inhibitory Fcγ receptor, FcγRIIb, are considered to increase ADCP activity [26]. Assessment of in vitro ADCP activity is often neglected during the course of mAb development as the assays can be tedious and labor and time intensive. Regardless, ADCP assays should be considered for CIT biotherapeutics containing an Fc domain including standard mAbs, Fc-engineered mAbs, ADCs, bispecific modalities, and other related antibody constructs, especially if the Fc domain is found to bind activating Fcγ receptors or if the intended MOA is tumor cell depletion. In vitro ADCP assays can inform in vivo MoA and may aid in lot biocompatibility assessments.

The classic format for in vitro ADCP assessment relies on PBMC-derived macrophages as effector cells. Monocytes from PBMCs are isolated and differentiated in culture to macrophages over several days [29]. Subsequently, macrophages are labeled with one fluorescent dye and target cells (tumor cells or other cells that express the target antigen) with another. Following a coincubation step of the labeled target and effector cells with therapeutic mAb, cells are harvested, and each cell population (macrophages, target cells and double positive cells) is distinguished by flow cytometry using absolute quantification of cell numbers or relative quantification with the help of beads to normalize results of decline in target cells [29]. However, it can be difficult to distinguish between target cells that are bound to macrophages versus those that have been phagocytosed using the flow cytometric method as both would be detected as double-positive cells. Additionally, it is also difficult to distinguish between macrophages that have phagocytosed one or multiple tumor cells. Confocal microscopy can be used to investigate the latter, but quantification with this technique is highly labor intensive [30]. A relatively recent development is the use of imaging flow cytometers, which combine capabilities of microscopy and flow cytometry in a single platform, which among other parameters can provide information on location of fluorescent dyes and co-localization of different fluorescent probes. Alternatively, target cells may be labeled with a pH-sensitive dye that only fluoresces in mature phagosomes to quantify frank phagocytosis [31]. As with the ADCC assay, effector-to-target cells ratios is an important factor in the optimization of the ADCP assay. Another important consideration is that effector cells cannot be pooled due to MHC restriction, and a battery of donors must be applied to reduce donor-specific variability.

Reporter gene bioluminescence assays have also been established to quantify ADCP that use engineered cell lines expressing FcγRIIa as effector cells [32, 33]. In this technique, target cells are cocultured with therapeutic mAb and FcγRIIa-expressing effector cells. The interaction between the Fc domain of the target cell-bound mAb and FcγRIIa results in FcγRIIa signaling in the effector cells and expression of the reporter gene whose bioluminescence can be quantified. Although this assay may be conducted more rapidly and efficiently than the classic ADCP assay, it is not a direct measure of ADCP activity. Therefore, it may be used as a screening tool in preliminary stages to be followed up by the classical ADCP assay to demonstrate true biological relevance.

3.3 CDC Assays

In addition to ADCC and ADCP activities, Fc domains have the further potential to initiate complement-dependent cell lysis known as CDC. CDC is an important part of the MoA of many tumor-depleting mAbs especially those targeting lymphomas [34]. CDC is initiated when cell surface target-bound mAbs recruit and bind to complement factor, C1q, via the Fc domain. This triggers the complement cascade resulting in the formation of a membrane attack complex (MAC) on the cell surface, which in turn results in target cell lysis. This process is known as the classical pathway of complement activation. The process of complement activation and cell lysis can result in inflammation and subsequent recruitment and stimulation of immune cells, which has the potential to mediate toxicity. In general, CDC effector function is considered high for human IgG1 and IgG3 isotypes, low for IgG2, and null for IgG4 [34, 35] and is tightly correlated with the Fc domain's ability to recruit and bind to C1q. Moreover, amino acid mutations introduced into Fc regions have been shown to modulate C1q binding, which could modify CDC activity [10, 15]. Regardless, CDC assays should be considered for all CIT biotherapeutics containing an Fc domain including standard mAbs, Fc-engineered mAbs, ADCs, bispecific modalities, and related molecules. As with the ADCC assay, the CDC assay is a standard endpoint included in the early in vitro assessment of these modalities to inform both MoA and mechanisms of potential immune-mediated toxicities.

Similar to the classical ADCC assay, the classical approach for the in vitro CDC activity also employs a radioactive methodology. In this assay, target cells (primary human cells or cell lines) expressing the antigen are loaded with ^{51}Cr and incubated with the therapeutic mAb at various concentrations and human serum comprising the components of the complement system. Cell death is determined by release of ^{51}Cr in the supernatant by scintillation counts [32]. As both ADCC and CDC assays can utilize ^{51}Cr-based methods and same target cells to determine cell death, the two assays are frequently conducted in parallel. While the radioactive method is still considered the most sensitive, there is a push to use nonradionuclides due to environmental and operator protection constraints [32]. More recently, newer nonradioactive methods have been developed to assess CDC activity which can be determined by the release of cellular components such as GAPDH and proteases or by use of fluorescent or luminescent probes, such as calcein and alamar blue. Flow cytometric CDC assays have also been developed wherein cell death is interrogated by staining with an intercalating dye to identify dead targets [36].

There are several key factors that can enhance or decrease CDC and should be considered in the assay development. Apart from the antibody isotype and amino acid substitutions that affect CDC, the expression level of antigens and complement regulators on target cells may also influence CDC activity. If target cells do not express sufficient level of target antigen for the mAb to bind and recruit complement, the complement activation process cannot be initiated and no CDC activity may be observed [37]. Secondly, some membrane-bound regulators such as CD46, CD55, and CD59 may regulate complement activation positively or negatively if expressed on the surface of target cells [38, 39]. Finally, as CDC activity is strongly correlated with C1q recruitment and binding, direct binding of the Fc to C1q should be assessed by plasmon resonance or immunoassays prior to conduct of the cell-based CDC assay. While the C1q binding assay does not directly predict CDC activity, it acts as a supplement to the in vitro CDC assay and provides additional mechanistic insight.

4 Effector Cell Activation/Target Cell Killing Assays

One of the most rapidly growing therapeutic platforms in CIT is synthetic antibodies to redirect effector immune cell population against target tumor cells [40, 41]. These cell-engaging antibodies typically have dual antigen specificity, binding to two unique antigens one on effector cells (e.g., CD3 on T-cells) and one cell surface target on tumor cells (e.g., CD19 on B-cells), and thereby facilitate cell-to-cell interactions (Fig. 2). This direct contact is sufficient to cause a rapid activation of effector cells to mediate cytotoxic responses via production of cytokines and other cytolytic proteins (e.g., perforin, granzyme) to kill target tumor cells. Compelling clinical efficacy has been demonstrated with blinatumomab, the first-generation CD3/CD19 bispecific T-cell redirecting antibody fragment for the treatment of relapsed or refractory B-cell precursor acute lymphoblastic leukemia. However, from the safety perspective, profound and rapid cytokine production has been observed with this class of molecules, which represents a potential

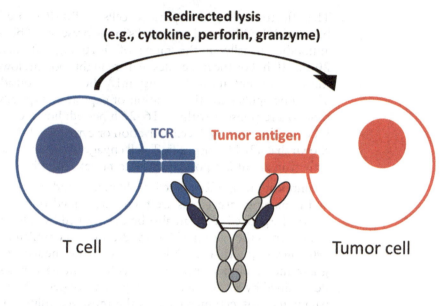

Fig. 2 Mechanism of action of T-cell redirecting bispecific antibody

of acute phase responses and often defines the maximum tolerated dose (MTD).

The MoA for this class of molecules is relatively straightforward. For example, engaging both antigens is sufficient to cause T-cell activation, which is independent of the intrinsic antigen-specific TCR recognition. Therefore, a battery of in vitro pharmacologic activity assessments including effector cell proliferation and/or activation, target cell cytotoxicity, and cytokine production can be effectively conducted in culture systems containing both effector and target cells at certain ratios. In general, T-cell bispecific antibodies are quite potent at inducing cytokine production and tumor cell killing in the picomolar range (pM) [42–44]. In contrast, no similar pharmacological effects are manifested when only effector or target cells are present in the culture.

The general procedure of effector cell activation/target cell killing in vitro assays includes three major steps:

1. The target cells are preincubated for 1–2 h with IgG (approximately 0.5 mg/mL) to reduce nonspecific binding to cell surface FcγRs by test molecules (this step may not be needed for molecules where effector function has been engineered out.) In vitro and in vivo studies have shown that nonspecific killing by nontargeting conjugates occurs if there is no preincubation or predosing with human IgG or mouse IgG2a [45]. Furthermore, effectorless Fc does not necessarily prevent nonspecific binding to Fc or binding to cell membranes. This step is crucial as it minimizes/eliminates the nonspecific activity-related signals in the assay readout.

2. The effector cells (either purified cells or PBMCs) are incubated with IgG-blocked target cells in the presence of bispecific antibodies usually in the range of 0–10 mg/mL between 20 and 40 h. For the molecules that are highly potent, lowering the test concentration (0–1 mg/mL) should be considered. The same applies for the duration of exposure. Typically, the time to a response is within a 16–24 h period; however, earlier is better if measuring T-cell activation or cytokines. It has been shown that a BiTE (bispecific T-cell engager) can have maximal activity in about 2 h postexposure to target and effector cells.

3. Quantification of effector cell proliferation and/or activation and remaining target cells are then determined by flow cytometry. Target cell lysis can also be determined by cell viability assay such as the calcein AM assay. Production of inflammatory cytokines (e.g., IL-1α, TNF-α) or cytotoxic mediators (e.g., granzyme B or perforin) in the culture media can also be determined by ELISA or a multiplex assay. In general, multiple parameter flow cytometry offers the most versatility for endpoint measurement including cell numbers, viability, activation, membrane and intracellular expression of proteins, apoptosis, cell cycle, etc. In contrast, cytotoxicity assays such as calcein AM, Cell Titerglo, or imaging assay (IncuCyte), etc., are usually single-parameter assays with the exception of IncuCyte, which measures kinetics, and they are especially useful in early screening to identify lead molecules in a high-throughput capacity.

A good example of such in vitro assessments is related to anti-CD20/CD3 T-cell-dependent bispecific antibody (TDB) published by Sun et al. [46]. In their experiments, target BJAB cells (human Burkitt lymphoma B-cell line) and PBMCs isolated from healthy donor (E:T ratio of 10:1) were incubated with various concentrations of CD20/CD3 TDB for 24 h. In addition, in order to further assess how the various T-cell subsets contribute to the observed cytotoxic effects, purified $CD8^+$ T-cells or $CD4^+$ T-cells were incubated with BJAB cells at various concentrations of CD20/CD3 TDB for 24 h (E:T ratio of 5:1). BJAB cell killing and T-cell activation assessed by $CD69^+CD25^+$ expression were measured. Granzyme B induction was also detected by intracellular flow cytometry staining, and perforin concentration in culture medium was measured by ELISA. In brief, the results demonstrated that T-cell activation by CD20/CD3 TDB was strictly dependent on the presence of $CD20^+$ target cells, and CD20/CD3 TDB mediates B-cell killing in a clear T-cell-dependent manner with $CD8^+$ T-cells appearing more potent in target cell killing.

As illustrated by this example, those in vitro assays are particularly useful for characterizing the MoA of the immune cell redirecting bispecific antibodies. However, it has also been recognized that there is a notable dependence on effector-to-target (E:T) cell ratio, target expression, target cell, incubation time, and the bioanalytical platform used.

4.1 Effect of Target Expression

In general, the EC_{50} for a T-cell-redirecting bispecific antibody is expected to be collectively determined by a number of factors including target receptor binding epitopes [47], the antibody architecture [48], and the sensitivity of target cells to effector cell-mediated cytotoxicity [46, 49]. An impact of target expression on EC_{50} values is well demonstrated by Laszlo et al. on AMG330, which is a bispecific T-cell engager targeting CD33 and CD3 [50]. The experiments were conducted to evaluate the cytotoxicity effects of AMG 330 on engineered OCI-AML3 target cells (similar sensitivity to T-cell-mediated cytotoxicity due to the same parental cell line origin) with various CD33 expression levels. The lysis of OCI-AML3 cells (500, 1500, 5000, or 10,000 CD33 receptors per cell) by T-cells from healthy donors at E:T ratios of 10:1 or 1:1 were evaluated in the presence of AMG 330 after 48 h of incubation. In brief, the results showed higher copy number of target receptor CD33 and the lower EC_{50} value achieved in this assay system while maximal cytotoxicity appears comparable across the cells, which was comparable between both E:T ratios.

4.2 Effect of Effector-to-Target Ratio

Studies in the literature have shown that an almost complete target cell lysis can be demonstrated in in vitro assays (usually ≥24 h incubation) within a reasonable range of E:T ratios [50, 51]. As exemplified by Patrick et al., an E:T ratio as low as 1:5 caused complete $CD19^+$ Nalm-6 lysis after 24 h incubation with CD3/CD19 bispecific single-chain antibody construct [51]. In our own experience, in the case of T-cell bispecifics, both purified $CD8^+$ T-cells with various E:T ratios and PBMCs (inherent E:T, generally 1:10–1:1.7) perform similarly in in vitro cytotoxicity assays and, thus, PBMCs are routinely used for simplicity.

4.3 Effect of Target Binding Affinity

Another factor essential to the target cell killing effect for bispecific antibodies is their target binding affinity. Multiples studies have shown that a good balance between the intrinsic potency and target binding affinity is critical for optimal PD effects [43, 52]. In general, a relatively high target binding affinity is needed to achieve maximal target cell killing if the intrinsic potency is low even for target cells with a high level of target receptor expression (i.e., 20,000 receptors per cell).

As exemplified by the work of Leong et al. on a CD3/CLL1 T-cell directing bispecific antibody, during the in vitro cytotoxicity assay [45], tumor cell growth suppression is dependent on affinity to CD3. In that study, three humanized antibodies with differential affinities to human CD3ε were tested: low (50 nM), high (0.5 nM), and very high (0.05 nM). Dose-dependent survival of AML cell line EOL-1 and percent of $CD8^+CD69^+CD25^+$ (activated CD8 T-cells)

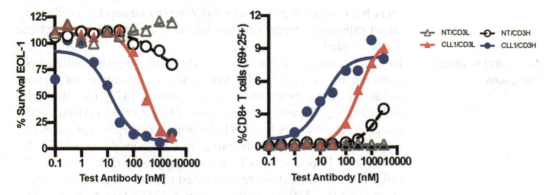

Fig. 3 Dose-dependent survival of AML cell line, EOL-1, and percent of activated CD8 effector (CD69$^+$CD25$^+$CD8$^+$) T-cells

effector cells were shown in Fig. 3. Similar results were observed for other AML cell lines. In brief, CD3/CLL1 TDB with a higher CD3 affinity demonstrated a lower EC$_{50}$ for tumor cell lysis and CD8 T-cell activation even though the maximal effects appear comparable. These in vitro effects seem to inversely correlate with the in vivo tolerability, namely the higher potency for the in vitro cell killing, the less tolerability in healthy naïve monkeys. With respect to the in vivo efficacy, the differences in mouse disease models were less profound between high and low CD3 affinity antibodies, likely due to prolonged drug exposure observed with lower-affinity CD3 TDBs. In brief, only the low-affinity CD3/CLL1 TDB was tolerated and associated with much less cytokine production in peripheral blood, which was sufficient to cause a target cell depletion in both peripheral blood and organ tissues (spleen and bone marrow).

Since most bispecific immunotherapeutics are highly potent both in vitro and in vivo, robust effector cell activation and associated cytokine production are often observed in animal studies, especially after the first dose when substantial target cells are present. When the target cells are reduced or even depleted, much less cytokine production is usually observed in subsequent doses. However, this first dose-related cytokine peak is recognized to be the key dose-limiting factor in animal studies. In our experience, tolerable dose levels of T-cell redirecting bispecific antibodies are usually in the low mg/kg range (i.e., ~1 mg/kg). Furthermore, it has been shown that the most common toxicities in both animals and patients were those related to cytokine release. Therefore, instead of using approaches such as receptor occupancy, highest nonseverely toxic dose (HNSTD), or no-observed adverse effect level (NOAEL), it is recommended that 10–30% pharmacologic activity (PA) using the corresponding EC values in the most sensitive in vitro assays is an acceptable approach for selecting the starting dose of the FIH clinical trials for T-cell redirecting bispecific antibodies [53]. In addition, a FIH dose higher than 30% PA (e.g., 50%) may also be acceptable when adequate justification is provided based on target biology and supporting data.

In summary, in vitro assays to assess effector cell activation and target cell killing provide great value for nonclinical assessment, especially for immune cell redirecting bispecific antibodies in two major aspects: (1) characterization of pharmacological effects and support lead antibody optimization and selection and (2) determination of dose response of pharmacologic activity and associated safety liabilities (e.g., cytokine production) to support a FIH starting dose selection. Several keys factors are known to have a major impact on the assay readout such as target expression, effector-to-target ratio, target binding affinity, etc. Therefore, it is critical to understand the differences between assay conditions to the expected situations in target patient population to optimize the use of in vitro assay data.

5 In Vitro Cytokine Release Assays (CRA)

The synthesis and release of cytokines from immune cells, under normal physiological conditions, are a highly regulated sequence of events [54]. A systemic inflammatory response can be triggered during infections or following the administration of novel therapeutic proteins. This large and rapid release of cytokines is termed as cytokine release syndrome (CRS). This term was first coined in the early 1990s when the anti-T-cell antibody muromonab-CD3 was introduced into the clinic as an immunosuppressive treatment for solid organ transplantation [55, 56]. Four putative molecular mechanisms by which a therapeutic mAb can cause the release of cytokines have been described. These include (1) direct activation and release of cytokines from cells bearing target antigen, (2) cross-linking by anti-IgG, activation and release of cytokines from cells bearing target antigen, (3) cross-linking by FcR, activation and release of cytokines from cells bearing target antigen, and (4) cross-linking FcR, activation and release of cytokines from cells bearing FcR [57, 58]. CRS is most likely mediated by one or more of these mechanisms operating simultaneously. The different mechanisms likely lead to distinct patterns of cytokine release, thereby leading to diverse clinical manifestations of CRS.

The most notable adverse event occurred in London in 2006 during the FIH phase I clinical trial of a novel superagonist anti-CD28 monoclonal antibody that activates T-cells, TGN1412. Six normal healthy volunteers were enrolled in the trial of TGN1412. Within 90 min after receiving a single intravenous dose of 0.1 mg/kg TGN1412, all six volunteers had a systemic inflammatory response characterized by an increase in levels of proinflammatory cytokines, which was accompanied by myalgias, headache, nausea, diarrhea, erythema, vasodilatation, and hypotension. They became critically ill with pulmonary infiltrates and lung injury, renal failure, and disseminated intravascular coagulation within 16 h of dosing. Severe and unexpected depletion of lymphocytes and monocytes occurred within 24 h after infusion. Additionally, two patients developed cardiovascular shock and acute respiratory distress syndrome. Thankfully, all six patients survived, but one had to have his fingers and toes amputated. This acute severe adverse event following TGN1412 administration was later identified as CRS, as increased levels of TNF-α, IFN-γ, and IL-2, IL-6, and IL-10 were detected in treated volunteers [59–61].

Due to this inherent liability that might exist for certain therapeutic antibodies, a science-based assessment of cytokine release as part of the Investigational New Drug (IND) package may be warranted [53, 62, 63]. Importantly, nonclinical toxicology species including the cynomolgus monkey, which is in most cases a species of choice for safety assessment of novel biologics, are considered to be poor predictors of lymphocyte stimulation, proliferation, and cytokine release in humans. In the case of TGN1412, cynomolgus monkeys were selected as the nonclinical toxicology species based on 100% homology in the binding site in the CD28 molecule and comparable expression of CD28 and T-cell subset ratios between humans and monkeys. However, studies have shown that $CD4^+$ effector cells in cynomolgus monkeys do not express CD28 and, therefore, cannot be stimulated by anti-CD28 agonists. However, high levels of CD28 are expressed in the $CD4^+$ memory T-cells in humans and, therefore, can be highly stimulated by TGN1412 [61, 64]. This inherent shortcoming in cynomolgus monkeys necessitated the development of novel in vitro assay formats for cytokine release hazard identification for biotherapeutics. Significant progress has been made in designing and developing methods for cytokine release as a result of the TGN1412 incident, and several cytokine release assays (CRAs) have been developed.

In 2009, the European Medicines Agency (EMA) organized a workshop to discuss in vitro CRA assays wherein it was concluded that while a specific assay could not be endorsed at that time, human in vitro CRAs are definitely useful in predicting the hazard associated with a biotherapeutic's potential to produce cytokine release in humans [65]. However, not all CRA formats can discriminate between biotherapeutics that induce mild or moderate cytokine release. Additionally, these assays do not determine a threshold at which cytokines released may be associated with serious adverse events in humans. Instead, the assays are designed as tools for hazard identification.

The CRAs generally are divided into the soluble- and solid-phase assay formats [66, 67]. In the soluble-phase CRA, the test antibody at various concentrations in a solution form is incubated with heparinized human whole blood or PBMCs. The number of donors for whole blood of PBMCs varies from 5 to 32. Other variation of the soluble-phase CRA assay utilizes human umbilical vein endothelial cells (HUVEC) cultured with reconstituted leucocytes, which consists of PBMC and granulocytes in 100% autologous platelet-poor plasma or uses PBMCs precultured at high density to acquire reactivity to TGN1412. In all these assay platforms, the plasma is collected following incubation for about 4–72 h, and various cytokines are subsequently measured. In 2007, a solid-phase CRA, which involves the coincubation of human peripheral blood mononuclear cells (PBMCs) with therapeutic antibodies that have been dry-coated onto a tissue culture plate, was shown to be predictive for the cytokine release potential of TGN1412. To avoid an edging effect observed in the dry-coat method wherein the edges of the well dry faster than the center, a wet coating method has gained popularity and has shown to be equally effective to the dry coating method. PBMCs from 6 to 20 donors are added to the coated culture plates. The supernatant is then collected followed by measurement of cytokines. In general, the solid-phase assay format in the presence of PBMCs elicits higher cytokine levels compared to other assay formats, and this is currently the most widely used assay format for cytokine assessment. Commonly tested cytokines in the CRA include granulocyte macrophage colony stimulating factor (GM-CSF), TNF-α, IFN-γ, IL-1β, IL-2, IL-4, IL-6, IL-8, IL-10, and IL-12. In a comparative study, Vessillier et al. have reported that the whole-blood CRA assays were the least sensitive for predicting cytokine release for TGN1412 as these required the largest group sizes. Conversely, the PBMC solid phase assay was most predictive for TGN1412 and muromonab CD3 (anti-CD3), which stimulates TGN1412-like cytokine release, as these required smaller group sizes. Conversely, the whole-blood CRA was far more sensitive than the PBMC solid-phase assay for alemtuzumab (anti-CD52), which stimulates FcγRI-mediated cytokine release. Further investigation revealed that removal of red blood cells from whole-blood assays permitted PBMC-like TGN1412 responses in a solid-phase CRA [68]. A retrospective analysis of 32 Investigational New Drug (IND) applications for immune checkpoint inhibitors submitted to the Food and Drug Administration showed that about half of the INDs conducted both the soluble- and solid-phase assays whereas the other half conducted only one platform assay [62]. Data from the CRAs were mostly used for hazard identification and in some cases for FIH dose selection. This analysis could not identify an assay platform that was the most sensitive or relevant in hazard identification as results varied depending on the product and the condition of use.

In summary, novel therapeutic proteins have the potential to cause cytokine release in clinical trials. Nonclinical toxicity species are generally not predictive of this hazard. A number of different

CRA formats are available to test for potential liabilities associated with novel biologics. Investigators should use a science-based approach while choosing the best assay format based on the biotherapeutic's MoA. Additionally, there might be scenarios where the biologic agent may have to be tested in multiple assay formats to appropriately characterize safety.

6 Tissue Cross-Reactivity Assays

Tissue cross-reactivity (TCR) studies are in vitro screening assays conducted with mAbs and related antibody-like biopharmaceuticals to identify off-target binding to unexpected epitopes. A secondary purpose of TCR studies is to identify on-target binding and tissue distribution that were not previously identified by other methodologies [69]. In general, a TCR study is recommended for any biotherapeutic containing a complementarity-determining region (CDR) [69, 70]. The ICH S6 guidance states that these studies are performed to support investigational new drug (IND) or clinical trial applications (CTAs) for the initial dosing of human subjects for all antibody-based therapeutics [71]. For biotherapeutics with more than one CDR, such as BiAbs or TDBs, there is no need to study individual binding components separately.

TCR studies use immunohistochemistry techniques to characterize binding to antigenic determinants in a broad panel of frozen human tissues using the mAb therapeutic as the primary reagent [69, 72] and require careful selection of appropriate positive and negative control tissues. Accurate interpretation of TCR staining requires a sound experimental IHC method and comparison of staining in investigational tissues with that in the positive and negative controls. There are three possible outcomes of the TCR study in any given tissue: negative staining, expected positive staining, and unexpected positive staining [72]. Since the study interpretation is heavily guided by unequivocal staining in the positive control tissue sample, the selection of the positive control is one of the main challenges to conduct a TCR study. Different laboratories and pathologists have a wide range of approaches, opinions, and criteria concerning what constitutes a satisfactory positive control. For tissue-bound targets, positive control is usually the normal or diseased organ tissues (e.g., tumors or inflamed tissues). For soluble targets which are not tissue bound, a positive control tissue may constitute an immortalized cell line overexpressing the target or tumor tissue [72]. Interpretation of TCR results requires professional judgment from a pathologist as well as robust peer review. An additional challenge to TCR studies is that therapeutic mAbs are generally not intended to be used as reagents for IHC. As a result, significant optimization is required in the course of establishing a reliable IHC methodology for TCR.

Since TCR studies are mainly aimed at identifying areas of unexpected staining to alert investigators of an in vivo target organ, they are generally conducted with a full panel of human tissues only. Assessment of TCR in a full panel of animal tissues is considered of limited value and is not recommended [71]. In certain cases, where no other method is available to ascertain pharmacologically relevant species, a comparison of the ex vivo patterns of staining between human and animal tissue panels may also be used to support the relevance of species evaluated in preclinical toxicity studies. More often, TCR studies may be conducted in a full or limited panel of tissues from nonclinical species to aid in the interpretation of toxicity study results. TCR studies are not recommended for assessing comparability of the test article as a result of manufacturing changes over the course of a development program or for determining tissue binding of surrogate mAbs used in efficacy studies [69].

While GLP-compliant TCR studies have become routine in the development of mAbs and other CDR-containing biotherapeutics, their value for the purpose of making safety assessment decisions by both industry and regulatory agencies has been questioned as experience with the assay has increased [4, 69, 72, 73]. It is important to remember that TCR study conditions in an ex vivo histologic tissue sample do not represent the complex tissue microenvironment, and the presence of ex vivo staining does not necessarily mean there will be any associated functional activity or toxicity in vivo. TCR studies also cannot detect subtle changes in critical quality attributes related to tissue fixation techniques. Therefore, a thorough understanding of the limitations of the method used is critical in extrapolating results of the TCR in assessing potential risk to humans. Based on results from a recent industry-wide survey, which included input from 17 pharmaceutical companies, it was apparent that different companies and scientists utilize and interpret the TCR assay in different ways [73]. The majority of companies did not use TCR data as the only basis for determining species selection for toxicity studies. It was also apparent that toxicities of only a small percent of molecules surveyed would have been missed in animal or clinical studies had TCR studies not been performed. The relevance and value of the TCR study is being continually assessed as experience in animals and human subjects accumulates.

6.1 Consideration for TCR with CIT Molecules

While a GLP-compliant TCR study is currently considered a standard component safety assessment of mAbs and related antibody-like biotherapeutics containing CDRs, careful consideration should be given to whether a TCR study is needed for these classes of molecules in the CIT field. A key consideration affecting the conduct of TCR for CITs is the intended patient population. If the intent is to dose only advanced cancer patients, i.e., those indications covered by the ICH S9 guidance [74], a TCR study may not be necessary. A recent clarification supplement to ICH S9 by an implementation working group states that TCR studies with biotherapeutics containing CDRs (mAbs, antibody-related products, ADCs, etc.) are of little utility and not needed for clinical dosing in indications covered by the S9 guidance unless there is a specific cause for concern [75].

Despite this guidance, TCR studies may still be needed in some cases for CIT biotherapeutics. As stated in the guidance itself, in cases where there are no pharmacologically relevant species, human TCR or alternative in vitro methods are still considered necessary to support FIH dosing [75]. Additionally, TCR assays may be conducted if the target expression and/or function in humans is not well understood if the CDR-containing molecular platform is first-in-class and if there are unexpected toxicity findings in nonclinical species. In all these cases, the TCR assay would be a supplementary endpoint supporting target RNA tissue expression profile, corroborating target organs of toxicity, or confirming lack of off-target interactions of novel platforms. A final consideration for conduct of TCR studies with CIT biotherapeutics is the intended patient population. In some cases, a biotherapeutic planned to be dosed only in advanced cancer indications covered by ICH S9 may be simultaneously or subsequently developed for dosing in patients without advanced cancer (for example, in the adjuvant setting, in noncancer autoimmune indications, etc.). In these cases, consistent with ICH S6, a TCR study in a full panel of human tissues may be necessary for FIH dosing in the absence of suitable or substantial clinical safety data. The conduct of TCR studies for CIT molecules, therefore, should follow a case-by-case approach taking into consideration tissue expression of the target, novelty of the platform, utility in explaining nonclinical findings, and planned patient population and indication.

7 Receptor Occupancy (RO) Assays

The measurement of the binding of a therapeutic biologic to its cellular target is also known as receptor occupancy (RO), and it has become increasingly important in the development of biotherapeutics. RO assays are commonly determined by flow cytometry mainly because these are able to effectively measure distinct cell subsets in heterogeneous populations, which is critical especially for oncology clinical trials. Data outputs include relative frequency (percent positive or absolute counts) or expression levels expressed as the fluorescence intensity. In practice, RO assays can be individualized to characterize different biological aspects such as internalization or shedding once a receptor is conjugated with the test agent. In the current scheme of drug development, RO assays are commonly used to support the assessment of the pharmacodynamic (PD) profile of a drug candidate. When combined with the pharmacokinetic (PK) profile, RO can provide valuable information pertaining to PK-PD relationships and thus support modeling whether given doses of a drug candidate at specific schedules lead to predicted levels of receptor occupancy and whether the receptor level is modulated on cells engaged by the therapeutic agent. All of this information eventually will help with the dose selection in either nonclinical animal toxicology studies or clinical trials in patients.

From a technical perspective, flow-cytometry-based RO assays generally use fresh blood samples, and assay readout can be potentially affected by a number of factors such as assay format, receptor expression level, study samples, and data normalization and reporting. For example, assay format is one of the most important components for RO assays, and it needs to be determined largely based on the MoA of each individual drug candidate. For antagonistic antibodies, the free assay format (Fig. 4a) is preferred as it is used to measure receptor availability for ligand binding by using a detection antibody conjugated with enzymes or fluorescent dyes. Alternatively, a drug-occupied receptor can be measured directly by using an antiidiotype detection antibody (Fig. 4b), and free receptor levels can be theoretically derived by PK-PD modeling. In contrast, in some cases it is important to have an actual measurement of all receptors, either free or bound (Fig. 4c). Another important factor that needs to be considered for designing the RO assay is the target expression level. In some cases, target receptors are expressed at low levels on circulating cells, especially in healthy individuals, and, therefore, the assay dynamic range may be too narrow to enable an accurate RO measurement. Possible solutions include either use of a high affinity antibody to enhance the signal intensity or ex vivo

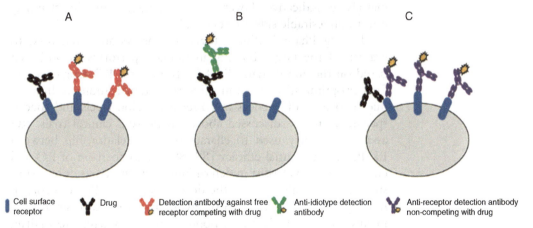

Fig. 4 Various assay formats for determining receptor occupancy. (**a**) Free receptor detection. (**b**) Occupied receptor detection. (**c**) Total receptor detection

stimulation to increase the target expression which, however, can present additional uncertainty in extrapolating those data to in vivo studies or to humans. Furthermore, the study population is another critical consideration as receptor expression levels could differ across animals of different origins or across patient populations. In general, an RO assay development should be conducted in animals of the same origin as the study population. For clinical studies, the RO assay development is generally carried out with blood from healthy subjects but should be evaluated in diseased samples prior to implementation, if possible.

7.1 Implementation

There has been a marked increase in the use of RO assays in many aspects of drug development at various stages. During the discovery stage of development, RO assays are conducted mainly to (1) evaluate the target engagement and its relationship to the observed efficacy in nonclinical disease model studies to assist in candidate screening [76–79], (2) select doses for animal toxicology studies coupled with PK modeling, and (3) determine the dosing regimen for initial clinical studies such as the starting dose selection for the first-in-human trials in combination with PK data [62, 80, 81]. From a safety perspective, RO is also frequently used to ensure a safe FIH starting dose especially when the expected safety liabilities are highly related to the exaggerated pharmacology and the MABEL-based approach is recommended. For example, a retrospective analysis of immune-activating biotherapeutics has been conducted to evaluate the approaches used to select the FIH starting dose [62]. The results showed that FIH doses based on 20–80% RO have acceptable toxicities for all immune-activating biotherapeutics examined (except T-cell engagers). It is important to note that 80% is below the RO that resulted in the cytokine storm with TGN1412 (anti-CD28 super agonistic mAb, 90% RO at the FIH dose). Furthermore, over chronic dosing, sustained maximum RO may also be indicative of overdosing or long-term binding that may lead to undesirable side effects [82, 83].

During Phase 1 clinical trials, RO assays are often used to evaluate if the target binding in humans performs as projected based on the animal data [84–87]. In general, RO assays require further optimization when moving from animal studies to humans due to species differences in target expression levels and species specific ligands as discussed above. In Phase 2 clinical trials, RO assays are largely used to characterize the relationship between PK-PD and potential efficacy [88, 89]. A comparison of RO and efficacy signals will assist in determining the efficacious dose/exposure range and also inform the dose selection for Phase 3 trials if moving forward. In Phase 3 or pivotal trials, RO assessments are mainly used to help define the population PK characteristics of drug candidates [81]. Several demographic factors such as age, weight, and gender, etc., may also be evaluated to determine if they could have any impact on the target binding and clinical efficacy [81, 82, 90]. Furthermore, RO assays can be used as comparators for new indications or the next generation molecules after the first approval.

Given the importance of RO assays in drug development, having a scientifically sound and validated assay is critical. Therefore, many supporting efforts such as assay development and validation must be conducted proactively to determine which method is most appropriate for RO assessment for each individual cancer immunotherapeutic [78, 81, 90]. A variety of factors could affect the RO assay outcome such as cell type, sample processing, and receptor internalization, etc., and thus need to be taken into consideration to optimize the assay development. When using nonclinical RO data to support human clinical study design, it is crucial to compare the receptor expression level between nonclinical species (mostly NHPs) and humans in order to determine if the animal data can be used to determine the dosing regimen in humans or any adjustment is needed.

Overall, in order to ensure reliable and high-quality data generation, interpretation, and application, careful RO assay design, appropriate data reporting, and well-integrated implementation are good practices for employing RO measurement in drug development.

8 Immunogenicity Assessment

Immunogenicity refers to the immune response elicited by the host toward a therapeutic protein. With the evolving complexity in therapeutic protein products, immunogenicity is often observed in the clinic. The generation of the immune response involves recognition of the therapeutic protein as foreign by the host's immune system. The foreign protein may then be taken up by antigen presenting cells (APCs), digested, and the smaller peptides presented on the cell surface by major histocompatibility complex (MHC) Class II. Recognition of these complexes by T-cells leads to their activation and in turn activates the B-cells. The B-cells then mature to plasma cells and produces antidrug antibodies (ADAs). The generation of ADAs (IgG, IgM, IgA, and/or IgE isotypes) by the body is a measure of the immunogenicity.

ADAs, in some instances, may act as neutralizing or nonneutralizing antibodies. A neutralizing antibody is one which binds to the biologic and inhibits its pharmacological function by preventing target binding. Whereas a nonneutralizing antibody, although binds to the biologic, does not affect target binding and, thereby, does not impact the drug's pharmacodynamic activity. However, a nonneutralizing antibody can lower the biologic's systemic exposure by increasing the clearance rate of drug, thereby finally leading to reduced clinical efficacy. Apart from affecting drug exposure and efficacy, ADAs can also cause toxicity by inducing hypersensitivity reactions [7]. Fundamental to any type of hypersensitivity reaction is the formation of an immune complex of an ADA. In type I hypersensitivity, IgE isotype ADAs are formed during an initial response, and upon repeat exposure to the biotherapeutic IgE-bound complexes bind and cross-link to Fc receptors on mast cells and basophils leading to degranulation, release of histamine, and manifestation of anaphylactic reaction, which in many cases may be fatal. Atypical anaphylaxis can also be triggered with IgG isotype ADA when such IgG bearing ICs formed after a second exposure cross-link Fcγ receptors on neutrophils, leading to activation and release of platelet-activating factor [91]. Type III hypersensitivity involves formation of large ADA/therapeutic protein complexes in the correct stoichiometry that do not get cleared but instead precipitate and deposit in tissues rich in filtering membranes made of fenestrated endothelium, such as kidneys, synovial membranes, and the choroid plexus. Further downstream tissue damage is mediated by complement fixation and activation or Fc-mediated inflammatory sequelae. This type of hypersensitivity is highly dependent on the ADA/drug ratio in the complex and might explain why it is relatively rare and variably seen despite the ease of forming immune complexes. ADA-mediated immune complex-induced toxicities have been reported in adlimumab-administered patients. Wherein, high antiadalimumab titer was associated with high risk of developing venous and arterial thromboembolism in three patients receiving adalimumab [92]. Due to the potential adverse effects of ADA on drug exposure, efficacy, and safety, assessment of ADA needs to be considered as part of the drug development program.

As expected, human therapeutic proteins often lead to immunogenicity when tested in nonclinical species. Therefore, nonclinical species are generally considered to be over predictive of human immunogenicity; moreover the immunogenicity observed in animal models may not translate to the clinic. Even though nonclinical immunogenicity assessment may not be valuable for clinical immunogenicity prediction, assessment of ADA in nonclinical studies might be necessary to determine the cause of unexpected exposure changes or to examine the mechanism of toxicities [93].

A risk-based approach is usually taken for immunogenicity assessment during clinical development, which includes in silico predictions during early stages of development followed by actual testing for immunogenicity using screening and validated methods [94]. As part of the in silico immunogenicity risk assessment, factors inherent to the biologic's structure and function, process of manufacturing, and the intended therapeutic use should be considered. These intrinsic factors have been systematically reviewed in a publication by Nadler et al. [95]. Briefly, from a structural standpoint, therapeutic proteins that are nonhumanized, possess CDRs that are divergent from germline, target immune cells, have low abundance of endogenous target proteins, and/or do not contain redundant target endogenous protein functions generally tend to have a high risk for immunogenicity. In contrast, a fully humanized sequence and immunosuppressive biotherapeutic tends to have a lower risk for immunogenicity. Manufacturing elements, including several design elements, could be perceived as foreign [96, 97]. A biologic with high immunogenicity risk might have unique conformational motifs present as incorrectly folded proteins and/or contaminated with host cell proteins. The route of administration is an important consideration for immunogenicity risk assessment wherein the risk for immunogenicity is considered to be the highest to lowest in the following order: inhalation → subcutaneous → intraperitoneal → intramuscular → intravenous. The immunogenicity risk might increase with increasing dose and frequency and duration of administration. Finally, patient factors, such as age and genetic predisposition; FcR polymorphisms; preexisting antibodies; and disease status, may also contribute to the risk of immunogenicity.

The US Food and Drug Administration (FDA), European Medicines Agency (EMA), and International Consortium of Harmonization (ICH) have published guidelines recommending evaluation of immunogenicity for therapeutic proteins along with guidance for assay development and validation during drug development [98, 99]. The guidance recommends developing screening assays during early stages of development followed by validated assays during later stages of development. Several laboratory methods are used for ADA assessment in nonclinical and clinical studies. In the past, high-pressure liquid chromatography (HPLC)-based methods and radioimmunoassay (RIA) were commonly used for ADA detection. Currently, safer and more high-throughput ELISA and electrochemiluminescence immunoassay (ECLIA) are preferred. Other technologies such as surface plasmon resonance (SPR) and bio-layer interferometry (BLI) are also employed for ADA assessment. These assays are used to detect ADAs with low affinity but have lower throughput compared to the immunoassays [100, 101].

In conclusion, novel therapeutic biologics have the potential to be immunogenic in the clinic, thereby producing ADAs and affecting the drug exposure and efficacy in addition to potentially causing life-threatening toxicities. Therefore, immunogenicity assessment during clinical development of novel biologics has been recommended by various regulatory agencies. The standard nonclinical species are generally not a quantitative or qualitative predictor of clinical immunogenicity, thus necessitating the use of in silico approaches and actual testing of clinical ADA during development.

9 Immune Function Assessments

Immune function assessments refer to endpoints conducted using blood and tissues collected within in vivo pharmacology or toxicity studies. Unlike standard in vitro endpoints, specialized immune function endpoints may require an additional stimulation or culture step to further assess drug effects. The main aim of these assays is to demonstrate or validate the MoA of the therapeutic which may not otherwise be observable using standard methods. In addition, immune function assays may also act as biomarkers of pharmacological activity within toxicity studies, especially in cases where the target is minimally or not expressed in nonclinical species. Immune function assays can further inform clinical biomarker monitoring plans and act as markers of efficacy. As cancer immunotherapeutics are aimed largely at enhancing antitumor immune responses, ex vivo assays for this class of drugs are specifically aimed at assessing immune enhancement, antigen specificity of immune activation, and whether the pharmacological immune activation may play a role in potential adverse events. The design and decision for inclusion of specialized ex vivo immune endpoints within in vivo studies should be guided by a thorough understanding of the CIT molecule's MoA and target expression level in the nonclinical species.

9.1 Antigen Challenge Models

Antigen challenge models were originally developed as comprehensive immune function assays for evaluating in vivo immunotoxicity potential of new drug candidates. The T-cell-dependent antibody response (TDAR) is the most widely used assay to assess immunotoxicity potential and is recommended by the ICH S8 guidance [102]. TDAR assays using keyhole limpet hemocyanin (KLH) as the challenge antigen have been widely reported [103, 104] and have been optimized to measure immunosuppressive effects of drug candidates on anti-KLH IgM and IgG antibody responses. Historically, immunotoxic potential of drugs referred largely to adverse effects on humoral or cellular immune function (i.e., immunosuppression) that may result in decreased host defense against infection and/or increased cancer risk [105]. By contrast, immunotoxic potential for cancer immunotherapeutics refers largely to their theoretical ability to enhance generalized (i.e., nontumor specific) immune responses as a consequence of exaggeration of pharmacology, which may result in inflammatory or autoimmune-type toxicities [106, 107]. Antigen challenge models previously validated to assess immunosuppression may, therefore, have to be modified to inquire immune enhancement potential. Challenge antigen doses have been historically titrated to elicit robust antigen-specific antibody responses so that immune suppression can be assessed. In order to assess potential for immune enhancement, a lower, suboptimal dose challenge antigen may be required.

In theory, immunization with KLH or other challenge antigens (e.g., hepatitis B surface antigen, tetanus toxoid, etc.) can be incorporated into nonclinical toxicity studies with CIT molecules across modalities and platforms to inquire immune modulation. In brief, the challenge antigen is administered to animals receiving the control and CIT drug in the context of repeat-dose toxicity studies either once (to inform effects on the primary immune responses) or 2–3 times (to inform effects on secondary or memory responses) depending on the length of the study and objectives [108]. At different time points throughout the study, serum and/or PBMC samples are collected and analyzed. Analysis may include the established in vitro measurement of challenge antigen-specific IgM or IgG antibodies and/or ex vivo T-cell responses upon restimulation of PBMCs with challenge antigen. Ex vivo T-cell response assays such as intracellular cytokine analysis by flow cytometry, ELISpot analysis for secreted cytokines, and cell proliferation assays may be more relevant endpoints for studying immune modulation by CIT drugs as a majority of these aim to enhance T-cell responses against tumors (Fig. 5).

Antigen challenges are used largely to identify potential immunomodulation-related hazards to humans [104]. Since many targets of CIT therapeutics are overexpressed in tumors, and either under expressed or not expressed in normal tissues, toxicity studies in healthy animals may not yield indications of immunomodulation-related risks. In these cases, antigen challenges provide a unique opportunity to activate the otherwise quiescent immune system, demonstrate pharmacological activity, make correlations between pharmacology and noted adverse drug-related effects, and inform potential risk to patients. However, data from

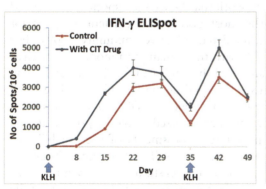 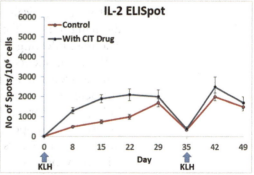

Fig. 5 Schematic representation of the enhancement of suboptimal KLH-induced secretion of IFN-γ and IL-2 by a CIT biotherapeutic in a cynomolgus monkey toxicity study upon ex vivo restimulation with KLH as measured by ELISpot analysis. KLH administration to assess primary (Day 0) and secondary (Day 35) immune responses is indicated by arrows

antigen challenge models and related endpoints should be interpreted with caution and within the context of other study findings including clinical pathology, histopathology and other immune function assessments. While a lack of effect does not imply lack of immunotoxic potential, a change in the functionality of immune cells is generally considered relevant to human subjects and should be carefully evaluated during the risk assessment phase [106, 108].

9.2 Viral Challenge Models

Like in many other fields of biological research, mouse models have been instrumental in demonstrating the therapeutic potential of some types of immunotherapies such as the first generation of mAbs against immune checkpoint regulators (e.g., CTLA-4, PD-1, or PD-L1). However, there are several notable limitations with current mouse models such as the differences between the mouse immune system and human tumor immunology (e.g., target expression). Therefore, developing alternative nonclinical models that bear more relevant immunobiology to CIT in humans would be of great value in supporting the development of next generation of immunotherapies. Since many immunotherapies aim to regulate the tumor-specific cytotoxic $CD8^+$ T-cells as one of the major effector cells involved in antitumor immunity, having a nonclinical model with predominantly $CD8^+$ T-cell-mediated immunity is key to clinically relevant assessments of pharmacological activity of drug candidates. One of the models commonly used for functional assessments of immunomodulators, especially immunosuppressants, is the model of T-cell dependent response (TDAR) to a specific antigen challenge. As mentioned above, the most common antigen that ICH S8 recommends for use in the TDAR model is KLH, which is known to be useful for functional assessments of $CD4^+$ T-cells and B-cells [108], and thus has limited utility for pharmacological assessments of immunotherapeutics mainly to modulate $CD8^+$ T-cell activity. In contrast, it has been understood from the vaccine-related work that DNA and especially viral-based vectors are typically the most potent and efficient antigens for inducing $CD8^+$ T-cell-mediated adaptive immunity [109]. For example, infecting NHPs with the simian immunodeficiency virus (SIV) components is an established animal model to study human immunodeficiency virus (HIV) pathogenesis, which has been well characterized to involve a major $CD8^+$ T-cell response [110]. Therefore, it has been recognized to have a good platform potential to support nonclinical assessments of the capability of immunotherapeutics to modulate $CD8^+$ T-cell immunity. Together with conventional tumor mouse model studies, the SIV model may provide additional support for a more integrated assessment of the potential for clinical efficacy of test molecules which may be more relevant to cancer. Furthermore, since many adverse events associated with cancer immunotherapies are mainly related to the pharmacology, this SIV NHP model could also provide a pharmacologically relevant assessment of safety liabilities for cancer immunotherapeutics.

Traditionally, SIV challenge-related studies are used for evaluating the protective efficacy of vaccine platforms, and thus usually involve challenging animals with replicating viruses (e.g., SIV-mac251 viral swarm) approximately 30 weeks after the initial vaccination, followed by monitoring viral load for several weeks (≥ 8–10 weeks) postinfection [110]. In total, the entire study takes about 40–50 weeks to complete which may not be ideal for a high-throughput screening of drug candidates. In order to make the assessment more efficient and practical, a modified TDAR-SIV model in NHPs is being tested and used preliminarily for screening immunotherapeutic candidates (Fig. 6). In general, NHPs are first vaccinated intramuscularly with nonreplicating adenovirus serotype 5 (Ad5) viral constructs encoding SIV proteins (Gag and Nef) to enable the tracking of known MHC-restricted $CD8^+$ T-cell epitopes. Following the immunization, vaccine-elicited T-cell responses and major T-cell subsets are measured by flow cytometry, including the use of peptide-loaded MHC class I tetramers (Nef RM9 peptide and Gag GW9 peptide), T-cell proliferation (e.g., Ki67) and activation (e.g., $CD69^+CD25^+$), and IFN-γ enzyme-linked immunospot (ELISPOT) assays. Immunotherapeutics are administered intravenously to NHPs immediately after vaccination to assess any effects on the CD8 T-cell response to the viral challenge.

As exemplified by Loffredo et al. [111], this TDAR-SIV model has been used to assess how the nonfucosylated anti-CTLA4 antibody could modulate the SIV-induced $CD8^+$ T-cell response in comparison to the first generation of anti-CTLA4 antibody ipilimumab. In brief, the results showed that a single dose of either antibody caused enhanced SIV antigen-specific $CD8^+$ T-cell responses over the vehicle at multiple time points with a greater

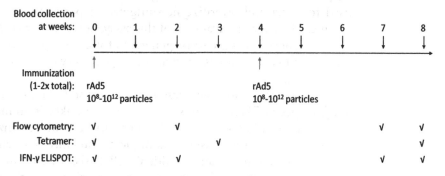

Fig. 6 Schematic of general methodology of modified TDAR-SIV model

effect demonstrated with the nonfucosylated anti-CTLA4 antibody. It provided a valuable nonclinical dataset to further characterize and support the utility of the modified NHP TDAR-SIV model for a clinically relevant pharmacological assessment of drug candidates targeting for $CD8^+$ T-cells.

9.3 MHC Tetramer Assay and ELISpot

The major histocompatibility complex (MHC) tetramer assay refers to a method that uses bioengineered MHC tetramer proteins with bound peptide antigen to quantify T-cells with receptors that are specific for that antigen. The tetramers are also labeled with a fluorophore allowing for direct visualization, quantification, and phenotypic characterization of antigen-specific T-cells using flow cytometry [112, 113]. Compared to traditional flow cytometric methods which assess numbers or percentages of immune cell subsets that may be difficult to detect, this method relies on the highly specific interaction between peptide-loaded MHCs and the corresponding T-cell receptor. Tetramers of both MHC class II and class I have been described which allow for visualization of $CD4^+$ and $CD8^+$ antigen-specific T-cell responses [113–115]. The MHC tetramer assay is particularly beneficial in the field of cancer vaccines which aim at enhancing T-cell responses or generating novel T-cell responses to tumor-specific antigens or cancer neoantigens and can be incorporated both in nonclinical as well as clinical studies. MHC tetramer assays help validate proof-of-concept (i.e., induction of antigen-specific T-cell responses), sensitivity, and robustness of T-cell responses, differentiation of $CD4^+$ vs. $CD8^+$ responses, and inform how responses wane over time. The demonstration of specific T-cell responses toward antigen(s) or neoantigen(s) encoded by cancer vaccines is uniquely useful in nonclinical studies because it acts as a marker of efficacy and allows for the use of healthy, nontumor bearing animals in pharmacology and toxicology studies to demonstrate proof of principle. In the clinical setting, information gathered using these assays can furthermore inform dosing frequencies and aid in fine tuning iterative prediction algorithms used to select and prioritize neoantigens for personalized cancer vaccines. A major disadvantage of this assay, however, is that MHC tetramer constructs have to be generated for each peptide of interest, which may be patient specific, and can be very costly and time consuming.

Another less costly and less specific method for monitoring antigen-specific T-cell responses is the enzyme-linked immunospot (ELISpot) assay [116, 117]. ELISpot employs a sandwich enzyme-linked immunosorbent assay technique to quantify frequency of cytokine secretion on an individual cell basis. In this method, PBMCs are cultured in the presence or absence of antigen(s) encoded by vaccines, and cytokine secretion from individual cells is measured upon binding to immobilized cytokine-specific antibodies in close proximity. The spots surrounding individual cells are counted with an ELISpot reader following conjugation with a detection antibody reagent. While ELISpot can measure multiple cytokine analytes simultaneously, IFN-γ is the most frequently measured cytokine by ELISpot and acts as a surrogate marker of antigen-specific CD8$^+$ T-cell responses [116]. Data have suggested that ex vivo IFN-γ ELISpot responses induced by cancer vaccines may correlate with disease outcome [118–120]. A major advantage of the ELISpot method is that it quantifies the number of responding cells, is very sensitive, and can be easily implemented as an ex vivo endpoint both in nonclinical and clinical studies. Disadvantages of this method are that it is time and labor intensive, detection cannot be coupled with further analysis, and identified cells cannot be isolated. Moreover, cytokine release alone may not always correlate with functional T-cell activity in vivo [116, 121].

9.4 Other Immunotoxicology Assessments

Nonclinical immunotoxicity assessments are usually conducted in a step-wise approach, starting with screening assays for a general understanding of any effects that drug candidates may have on the immune system. Immunological parameters/assays typically included in the first tier of assessments are (1) immunophenotyping of T-, B-, and NK cells to determine any changes with the major immune cell population in terms of cell counts (absolute or relative percentage), proliferation (e.g., Ki67) and/or activation (e.g., CD25$^+$/CD69$^+$), (2) of serum cytokine or immunoglobulin levels indicative of any immune activation or suppression, (3) clinical pathology to detect any inflammatory signals such as CRP increase, (4) histopathology to determine any immune-related changes in organ tissues such as immune cell trafficking-related tissue injury, etc. In many cases, those assays are sufficient to provide a solid nonclinical assessment of immunotoxicity potential for cancer immunotherapeutics and thus support advancing the drug candidates to the next development stage. In some cases, additional studies are warranted to further characterize the findings observed in the first tier of assessments such as any impact on the immune cell functions (e.g., ability of CD8$^+$ T-cells or NK cells to kill target cells) and thus to better understand the biological significance of those nonclinical findings. Moreover, based on the general MoA of immunotherapeutics to enhance the immune activity, studies to determine any potential toxicity in the context of vaccine immunization are another functional immunotoxicity assessment generally recommended for cancer immunotherapeutics [23].

For example, there is a concern that blocking the PD-1/PD-L1 pathway may exacerbate immune response to primary or recall antigen stimulation. Based on that consideration, patients who are vaccinated or revaccinated while undergoing treatment with anti-PD-1 or anti-PD-L1 monoclonal antibodies may experience an enhanced immune response resulting in vaccine-related toxicity.

Table 2
Other assays for functional immunotoxicological assessments

Immunotoxicological assessment	Endpoint and/or methodology	Reference
Immune cell proliferation and/or activation	Proliferation marker(s) by flow cytometry Activation marker(s) by flow cytometry	[122–124]
Cytokine/chemokine	Profiling by ELISA or multiplex Release by flow cytometry	[65, 125]
Complement system	Complement assays	[126–128]
Neutrophils and macrophages	Oxidative burst activity Phagocytosis Migration	[128–131]
NK and CD8$^+$ T-cell cytotoxicity	Target cell killing by gamma counter, or flow cytometry	[130, 132]

Therefore, performing an animal study to characterize the magnitude, kinetics, and resolution of the immune response following repeated vaccination (recall challenge) in anti-PD-1 or anti-PD-L1-treated animals versus control-treated animals would be of great value. As summarized in the ICH S8 guidance [102], the TDAR model has been identified in a regulatory context as a main functional test of immunotoxicity. However, it has been mainly used for immunosuppressants. As explained above, the revised TDAR model to KLH antigen challenge represents a good platform to evaluate if cancer immunotherapies could lead to an enhanced immune response to KLH immunization. In addition to the TDAR assay, additional functional assays, especially in vitro assays as summarized below (Table 2), can be used as follow-up studies to the first tier of nonclinical immunotoxicity assessments. Collectively, these assays provide an integrated assessment of immunotoxicity potential and its potential relevance to patients.

References

1. Decker WK, da Silva RF, Sanabria MH, Angelo LS, Guimaraes F, Burt BM, Kheradmand F, Paust S (2017) Cancer immunotherapy: historical perspective of a clinical revolution and emerging preclinical animal models. Front Immunol 8:829. https://doi.org/10.3389/fimmu.2017.00829
2. Ventola CL (2017) Cancer immunotherapy, part 3: challenges and future trends. P T 42 (8):514–521
3. Abdel-Wahab N, Alshawa A, Suarez-Almazor ME (2017) Adverse events in cancer immunotherapy. Adv Exp Med Biol 995:155–174. https://doi.org/10.1007/978-3-319-53156-4_8
4. Brennan FR, Kiessling A (2017) In vitro assays supporting the safety assessment of immunomodulatory monoclonal antibodies. Toxicol In Vitro 45(Pt 3):296–308. https://doi.org/10.1016/j.tiv.2017.02.025
5. Glass TR, Ohmura N, Saiki H (2007) Least detectable concentration and dynamic range of three immunoassay systems using the same antibody. Anal Chem 79(5):1954–1960. https://doi.org/10.1021/ac061288z
6. Bee C, Abdiche YN, Stone DM, Collier S, Lindquist KC, Pinkerton AC, Pons J, Rajpal

A (2012) Exploring the dynamic range of the kinetic exclusion assay in characterizing antigen-antibody interactions. PLoS One 7(4):e36261. https://doi.org/10.1371/journal.pone.0036261

7. Hansel TT, Kropshofer H, Singer T, Mitchell JA, George AJ (2010) The safety and side effects of monoclonal antibodies. Nat Rev Drug Discov 9(4):325–338. https://doi.org/10.1038/nrd3003

8. Beck A, Wagner-Rousset E, Ayoub D, Van Dorsselaer A, Sanglier-Cianferani S (2013) Characterization of therapeutic antibodies and related products. Anal Chem 85(2):715–736. https://doi.org/10.1021/ac3032355

9. Bruhns P, Iannascoli B, England P, Mancardi DA, Fernandez N, Jorieux S, Daeron M (2009) Specificity and affinity of human Fcgamma receptors and their polymorphic variants for human IgG subclasses. Blood 113(16):3716–3725. https://doi.org/10.1182/blood-2008-09-179754

10. Wang X, Mathieu M, Brezski RJ (2018) IgG Fc engineering to modulate antibody effector functions. Protein Cell 9(1):63–73. https://doi.org/10.1007/s13238-017-0473-8

11. Brezski RJ, Kinder M, Grugan KD, Soring KL, Carton J, Greenplate AR, Petley T, Capaldi D, Brosnan K, Emmell E, Watson S, Jordan RE (2014) A monoclonal antibody against hinge-cleaved IgG restores effector function to proteolytically-inactivated IgGs in vitro and in vivo. MAbs 6(5):1265–1273. https://doi.org/10.4161/mabs.29825

12. Tao MH, Smith RI, Morrison SL (1993) Structural features of human immunoglobulin G that determine isotype-specific differences in complement activation. J Exp Med 178(2):661–667

13. Wang Q, Chung CY, Chough S, Betenbaugh MJ (2018) Antibody glycoengineering strategies in mammalian cells. Biotechnol Bioeng 115(6):1378–1393. https://doi.org/10.1002/bit.26567

14. Beck A, Reichert JM (2012) Marketing approval of mogamulizumab: a triumph for glyco-engineering. MAbs 4(4):419–425. https://doi.org/10.4161/mabs.20996

15. Kellner C, Otte A, Cappuzzello E, Klausz K, Peipp M (2017) Modulating cytotoxic effector functions by Fc engineering to improve cancer therapy. Transfus Med Hemother 44(5):327–336. https://doi.org/10.1159/000479980

16. Junttila TT, Parsons K, Olsson C, Lu Y, Xin Y, Theriault J, Crocker L, Pabonan O, Baginski T, Meng G, Totpal K, Kelley RF, Sliwkowski MX (2010) Superior in vivo efficacy of afucosylated trastuzumab in the treatment of HER2-amplified breast cancer. Cancer Res 70(11):4481–4489. https://doi.org/10.1158/0008-5472.CAN-09-3704

17. Niwa R, Shoji-Hosaka E, Sakurada M, Shinkawa T, Uchida K, Nakamura K, Matsushima K, Ueda R, Hanai N, Shitara K (2004) Defucosylated chimeric anti-CC chemokine receptor 4 IgG1 with enhanced antibody-dependent cellular cytotoxicity shows potent therapeutic activity to T-cell leukemia and lymphoma. Cancer Res 64(6):2127–2133

18. Schlothauer T, Herter S, Koller CF, Grau-Richards S, Steinhart V, Spick C, Kubbies M, Klein C, Umana P, Mossner E (2016) Novel human IgG1 and IgG4 Fc-engineered antibodies with completely abolished immune effector functions. Protein Eng Des Sel 29(10):457–466. https://doi.org/10.1093/protein/gzw040

19. Overdijk MB, Verploegen S, Bogels M, van Egmond M, Lammerts van Bueren JJ, Mutis T, Groen RW, Breij E, Martens AC, Bleeker WK, Parren PW (2015) Antibody-mediated phagocytosis contributes to the anti-tumor activity of the therapeutic antibody daratumumab in lymphoma and multiple myeloma. MAbs 7(2):311–321. https://doi.org/10.1080/19420862.2015.1007813

20. Gómez Román VR, Murray JC, Weiner LM (2014) Antibody-dependent cellular cytoxicity (ADCC). In: Ackerman A, Nimmerjahn F (eds) Antibody Fc: linking adaptive and innate immunity, 1st edn. Elsevier Inc, San Diego, CA

21. Bakema JE, van Egmond M (2014) Fc receptor-dependent mechanisms of monoclonal antibody therapy of cancer. Curr Top Microbiol Immunol 382:373–392. https://doi.org/10.1007/978-3-319-07911-0_17

22. Zamai L, Ponti C, Mirandola P, Gobbi G, Papa S, Galeotti L, Cocco L, Vitale M (2007) NK cells and cancer. J Immunol 178(7):4011–4016

23. Nelson DL, Kurman CC, Serbousek DE (1993) 51Cr release assay of antibody-dependent cell-mediated cytotoxicity (ADCC). Curr Protoc Immunol 8(1):7.27.21–27.27.28

24. Parekh BS, Berger E, Sibley S, Cahya S, Xiao L, LaCerte MA, Vaillancourt P, Wooden S, Gately D (2012) Development and validation of an antibody-dependent cell-mediated cytotoxicity-reporter gene

assay. MAbs 4(3):310–318. https://doi.org/10.4161/mabs.19873
25. Zaritskaya L, Shurin MR, Sayers TJ, Malyguine AM (2010) New flow cytometric assays for monitoring cell-mediated cytotoxicity. Expert Rev Vaccines 9(6):601–616. https://doi.org/10.1586/erv.10.49
26. Gul N, van Egmond M (2015) Antibody-dependent phagocytosis of tumor cells by macrophages: a potent effector mechanism of monoclonal antibody therapy of cancer. Cancer Res 75(23):5008–5013. https://doi.org/10.1158/0008-5472.CAN-15-1330
27. Braster R, O'Toole T, van Egmond M (2014) Myeloid cells as effector cells for monoclonal antibody therapy of cancer. Methods 65(1):28–37. https://doi.org/10.1016/j.ymeth.2013.06.020
28. Kurdi AT, Glavey SV, Bezman NA, Jhatakia A, Guerriero JL, Manier S, Moschetta M, Mishima Y, Roccaro A, Detappe A, Liu CJ, Sacco A, Huynh D, Tai YT, Robbins MD, Azzi J, Ghobrial IM (2018) Antibody-dependent cellular phagocytosis by macrophages is a novel mechanism of action of elotuzumab. Mol Cancer Ther 17(7):1454–1463. https://doi.org/10.1158/1535-7163.MCT-17-0998
29. Su S, Zhao J, Xing Y, Zhang X, Liu J, Ouyang Q, Chen J, Su F, Liu Q, Song E (2018) Immune checkpoint inhibition overcomes ADCP-induced immunosuppression by macrophages. Cell 175(2):442–457 e423. https://doi.org/10.1016/j.cell.2018.09.007
30. Watanabe M, Wallace PK, Keler T, Deo YM, Akewanlop C, Hayes DF (1999) Antibody dependent cellular phagocytosis (ADCP) and antibody dependent cellular cytotoxicity (ADCC) of breast cancer cells mediated by bispecific antibody, MDX-210. Breast Cancer Res Treat 53(3):199–207
31. Velmurugan R, Ramakrishnan S, Kim M, Ober RJ, Ward ES (2018) Phagocytosis of antibody-opsonized tumor cells leads to the formation of a discrete vacuolar compartment in macrophages. Traffic 19(4):273–284. https://doi.org/10.1111/tra.12552
32. Rossignol A, Bonnaudet V, Clemenceau B, Vie H, Bretaudeau L (2017) A high-performance, non-radioactive potency assay for measuring cytotoxicity: a full substitute of the chromium-release assay targeting the regulatory-compliance objective. MAbs 9(3):521–535. https://doi.org/10.1080/19420862.2017.1286435
33. Zhang D, Whitaker B, Derebe MG, Chiu ML (2018) FcgammaRII-binding Centyrins mediate agonism and antibody-dependent cellular phagocytosis when fused to an anti-OX40 antibody. MAbs 10(3):463–475. https://doi.org/10.1080/19420862.2018.1424611
34. Rogers LM, Veeramani S, Weiner GJ (2014) Complement in monoclonal antibody therapy of cancer. Immunol Res 59(1–3):203–210. https://doi.org/10.1007/s12026-014-8542-z
35. Redpath S, Michaelsen T, Sandlie I, Clark MR (1998) Activation of complement by human IgG1 and human IgG3 antibodies against the human leucocyte antigen CD52. Immunology 93(4):595–600
36. Gillissen MA, Yasuda E, de Jong G, Levie SE, Go D, Spits H, van Helden PM, Hazenberg MD (2016) The modified FACS calcein AM retention assay: a high throughput flow cytometer based method to measure cytotoxicity. J Immunol Methods 434:16–23. https://doi.org/10.1016/j.jim.2016.04.002
37. Loeff FC, van Egmond HME, Nijmeijer BA, Falkenburg JHF, Halkes CJ, Jedema I (2017) Complement-dependent cytotoxicity induced by therapeutic antibodies in B-cell acute lymphoblastic leukemia is dictated by target antigen expression levels and augmented by loss of membrane-bound complement inhibitors. Leuk Lymphoma 58(9):1–14. https://doi.org/10.1080/10428194.2017.1281411
38. Geis N, Zell S, Rutz R, Li W, Giese T, Mamidi S, Schultz S, Kirschfink M (2010) Inhibition of membrane complement inhibitor expression (CD46, CD55, CD59) by siRNA sensitizes tumor cells to complement attack in vitro. Curr Cancer Drug Targets 10(8):922–931
39. Kesselring R, Thiel A, Pries R, Fichtner-Feigl S, Brunner S, Seidel P, Bruchhage KL, Wollenberg B (2014) The complement receptors CD46, CD55 and CD59 are regulated by the tumour microenvironment of head and neck cancer to facilitate escape of complement attack. Eur J Cancer 50(12):2152–2161. https://doi.org/10.1016/j.ejca.2014.05.005
40. Ellerman D (2018) Bispecific T-cell engagers: towards understanding variables influencing the in vitro potency and tumor selectivity and their modulation to enhance their efficacy and safety. Methods. https://doi.org/10.1016/j.ymeth.2018.10.026
41. Yu L, Wang J (2019) T cell-redirecting bispecific antibodies in cancer immunotherapy: recent advances. J Cancer Res Clin Oncol 145(4):941–956. https://doi.org/10.1007/s00432-019-02867-6

42. Dreier T, Lorenczewski G, Brandl C, Hoffmann P, Syring U, Hanakam F, Kufer P, Riethmuller G, Bargou R, Baeuerle PA (2002) Extremely potent, rapid and costimulation-independent cytotoxic T-cell response against lymphoma cells catalyzed by a single-chain bispecific antibody. Int J Cancer 100(6):690–697. https://doi.org/10.1002/ijc.10557

43. Durben M, Schmiedel D, Hofmann M, Vogt F, Nubling T, Pyz E, Buhring HJ, Rammensee HG, Salih HR, Grosse-Hovest L, Jung G (2015) Characterization of a bispecific FLT3 X CD3 antibody in an improved, recombinant format for the treatment of leukemia. Mol Ther 23(4):648–655. https://doi.org/10.1038/mt.2015.2

44. Krupka C, Kufer P, Kischel R, Zugmaier G, Lichtenegger FS, Kohnke T, Vick B, Jeremias I, Metzeler KH, Altmann T, Schneider S, Fiegl M, Spiekermann K, Bauerle PA, Hiddemann W, Riethmuller G, Subklewe M (2016) Blockade of the PD-1/PD-L1 axis augments lysis of AML cells by the CD33/CD3 BiTE antibody construct AMG 330: reversing a T-cell-induced immune escape mechanism. Leukemia 30(2):484–491. https://doi.org/10.1038/leu.2015.214

45. Leong SR, Sukumaran S, Hristopoulos M, Totpal K, Stainton S, Lu E, Wong A, Tam L, Newman R, Vuillemenot BR, Ellerman D, Gu C, Mathieu M, Dennis MS, Nguyen A, Zheng B, Zhang C, Lee G, Chu YW, Prell RA, Lin K, Laing ST, Polson AG (2017) An anti-CD3/anti-CLL-1 bispecific antibody for the treatment of acute myeloid leukemia. Blood 129(5):609–618. https://doi.org/10.1182/blood-2016-08-735365

46. Sun LL, Ellerman D, Mathieu M, Hristopoulos M, Chen X, Li Y, Yan X, Clark R, Reyes A, Stefanich E, Mai E, Young J, Johnson C, Huseni M, Wang X, Chen Y, Wang P, Wang H, Dybdal N, Chu YW, Chiorazzi N, Scheer JM, Junttila T, Totpal K, Dennis MS, Ebens AJ (2015) Anti-CD20/CD3 T cell-dependent bispecific antibody for the treatment of B cell malignancies. Sci Transl Med 7(287):287ra270. https://doi.org/10.1126/scitranslmed.aaa4802

47. Li J, Stagg NJ, Johnston J, Harris MJ, Menzies SA, DiCara D, Clark V, Hristopoulos M, Cook R, Slaga D, Nakamura R, McCarty L, Sukumaran S, Luis E, Ye Z, Wu TD, Sumiyoshi T, Danilenko D, Lee GY, Totpal K, Ellerman D, Hotzel I, James JR, Junttila TT (2017) Membrane-proximal epitope facilitates efficient T cell synapse formation by anti-FcRH5/CD3 and is a requirement for myeloma cell killing. Cancer Cell 31(3):383–395. https://doi.org/10.1016/j.ccell.2017.02.001

48. Gavrilyuk JI, Wuellner U, Salahuddin S, Goswami RK, Sinha SC, Barbas CF 3rd (2009) An efficient chemical approach to bispecific antibodies and antibodies of high valency. Bioorg Med Chem Lett 19(14):3716–3720. https://doi.org/10.1016/j.bmcl.2009.05.047

49. d'Argouges S, Wissing S, Brandl C, Prang N, Lutterbuese R, Kozhich A, Suzich J, Locher M, Kiener P, Kufer P, Hofmeister R, Baeuerle PA, Bargou RC (2009) Combination of rituximab with blinatumomab (MT103/MEDI-538), a T cell-engaging CD19-/CD3-bispecific antibody, for highly efficient lysis of human B lymphoma cells. Leuk Res 33(3):465–473. https://doi.org/10.1016/j.leukres.2008.08.025

50. Laszlo GS, Gudgeon CJ, Harrington KH, Dell'Aringa J, Newhall KJ, Means GD, Sinclair AM, Kischel R, Frankel SR, Walter RB (2014) Cellular determinants for preclinical activity of a novel CD33/CD3 bispecific T-cell engager (BiTE) antibody, AMG 330, against human AML. Blood 123(4):554–561. https://doi.org/10.1182/blood-2013-09-527044

51. Hoffmann P, Hofmeister R, Brischwein K, Brandl C, Crommer S, Bargou R, Itin C, Prang N, Baeuerle PA (2005) Serial killing of tumor cells by cytotoxic T cells redirected with a CD19-/CD3-bispecific single-chain antibody construct. Int J Cancer 115(1):98–104. https://doi.org/10.1002/ijc.20908

52. Ellerman D (2019) Bispecific T-cell engagers: towards understanding variables influencing the in vitro potency and tumor selectivity and their modulation to enhance their efficacy and safety. Methods 154:102–117. https://doi.org/10.1016/j.ymeth.2018.10.026

53. Saber H, Del Valle P, Ricks TK, Leighton JK (2017) An FDA oncology analysis of CD3 bispecific constructs and first-in-human dose selection. Regul Toxicol Pharmacol 90:144–152. https://doi.org/10.1016/j.yrtph.2017.09.001

54. Tisoncik JR, Korth MJ, Simmons CP, Farrar J, Martin TR, Katze MG (2012) Into the eye of the cytokine storm. Microbiol Mol Biol Rev 76(1):16–32. https://doi.org/10.1128/MMBR.05015-11

55. Charpentier B, Hiesse C, Ferran C, Lantz O, Fries D, Bach JF, Chatenoud L (1991) Acute clinical syndrome associated with OKT3 administration. Prevention by single injection

of an anti-human TNF monoclonal antibody. Presse Med 20(40):2009–2011

56. Sgro C (1995) Side-effects of a monoclonal antibody, muromonab CD3/orthoclone OKT3: bibliographic review. Toxicology 105 (1):23–29

57. Findlay L, Eastwood D, Stebbings R, Sharp G, Mistry Y, Ball C, Hood J, Thorpe R, Poole S (2010) Improved in vitro methods to predict the in vivo toxicity in man of therapeutic monoclonal antibodies including TGN1412. J Immunol Methods 352 (1–2):1–12. https://doi.org/10.1016/j.jim. 2009.10.013

58. Bugelski PJ, Achuthanandam R, Capocasale RJ, Treacy G, Bouman-Thio E (2009) Monoclonal antibody-induced cytokine-release syndrome. Expert Rev Clin Immunol 5 (5):499–521. https://doi.org/10.1586/eci. 09.31

59. Suntharalingam G, Perry MR, Ward S, Brett SJ, Castello-Cortes A, Brunner MD, Panoskaltsis N (2006) Cytokine storm in a phase 1 trial of the anti-CD28 monoclonal antibody TGN1412. N Engl J Med 355 (10):1018–1028. https://doi.org/10.1056/NEJMoa063842

60. Stebbings R, Findlay L, Edwards C, Eastwood D, Bird C, North D, Mistry Y, Dilger P, Liefooghe E, Cludts I, Fox B, Tarrant G, Robinson J, Meager T, Dolman C, Thorpe SJ, Bristow A, Wadhwa M, Thorpe R, Poole S (2007) "Cytokine storm" in the phase I trial of monoclonal antibody TGN1412: better understanding the causes to improve preclinical testing of immunotherapeutics. J Immunol 179(5):3325-3331

61. Danilenko DM, Wang H (2012) The yin and yang of immunomodulatory biologics: assessing the delicate balance between benefit and risk. Toxicol Pathol 40(2):272–287. https://doi.org/10.1177/0192623311430237

62. Saber H, Gudi R, Manning M, Wearne E, Leighton JK (2016) An FDA oncology analysis of immune activating products and first-in-human dose selection. Regul Toxicol Pharmacol 81:448–456. https://doi.org/10.1016/j.yrtph.2016.10.002

63. Saber H, Leighton JK (2015) An FDA oncology analysis of antibody-drug conjugates. Regul Toxicol Pharmacol 71(3):444–452. https://doi.org/10.1016/j.yrtph.2015.01.014

64. Eastwood D, Findlay L, Poole S, Bird C, Wadhwa M, Moore M, Burns C, Thorpe R, Stebbings R (2010) Monoclonal antibody TGN1412 trial failure explained by species differences in CD28 expression on CD4+ effector memory T-cells. Br J Pharmacol 161 (3):512–526. https://doi.org/10.1111/j.1476-5381.2010.00922.x

65. Vidal JM, Kawabata TT, Thorpe R, Silva-Lima B, Cederbrant K, Poole S, Mueller-Berghaus J, Pallardy M, Van der Laan JW (2010) In vitro cytokine release assays for predicting cytokine release syndrome: the current state-of-the-science. Report of a European Medicines Agency Workshop. Cytokine 51 (2):213–215. https://doi.org/10.1016/j.cyto.2010.04.008

66. Finco D, Grimaldi C, Fort M, Walker M, Kiessling A, Wolf B, Salcedo T, Faggioni R, Schneider A, Ibraghimov A, Scesney S, Serna D, Prell R, Stebbings R, Narayanan PK (2014) Cytokine release assays: current practices and future directions. Cytokine 66 (2):143–155. https://doi.org/10.1016/j.cyto.2013.12.009

67. Grimaldi C, Finco D, Fort MM, Gliddon D, Harper K, Helms WS, Mitchell JA, O'Lone R, Parish ST, Piche MS, Reed DM, Reichmann G, Ryan PC, Stebbings R, Walker M (2016) Cytokine release: a workshop proceedings on the state-of-the-science, current challenges and future directions. Cytokine 85:101–108. https://doi.org/10.1016/j.cyto.2016.06.006

68. Vessillier S, Eastwood D, Fox B, Sathish J, Sethu S, Dougall T, Thorpe SJ, Thorpe R, Stebbings R (2015) Cytokine release assays for the prediction of therapeutic mAb safety in first-in man trials--Whole blood cytokine release assays are poorly predictive for TGN1412 cytokine storm. J Immunol Methods 424:43–52. https://doi.org/10.1016/j.jim.2015.04.020

69. Leach MW, Halpern WG, Johnson CW, Rojko JL, MacLachlan TK, Chan CM, Galbreath EJ, Ndifor AM, Blanset DL, Polack E, Cavagnaro JA (2010) Use of tissue cross-reactivity studies in the development of antibody-based biopharmaceuticals: history, experience, methodology, and future directions. Toxicol Pathol 38(7):1138–1166. https://doi.org/10.1177/0192623310382559

70. Vugmeyster Y, Xu X, Theil FP, Khawli LA, Leach MW (2012) Pharmacokinetics and toxicology of therapeutic proteins: advances and challenges. World J Biol Chem 3(4):73–92. https://doi.org/10.4331/wjbc.v3.i4.73

71. (2011) ICH S6(R1): preclinical safety evaluation of biotechnology-derived pharmaceuticals. http://www.ich.org

72. Geoly FJ (2014) Regulatory forum opinion piece*: tissue cross-reactivity studies: what constitutes an adequate positive control and how do we report positive staining? Toxicol Pathol 42(6):954–956. https://doi.org/10.1177/0192623313495336
73. Bussiere JL, Leach MW, Price KD, Mounho BJ, Lightfoot-Dunn R (2011) Survey results on the use of the tissue cross-reactivity immunohistochemistry assay. Regul Toxicol Pharmacol 59(3):493–502. https://doi.org/10.1016/j.yrtph.2010.09.017
74. (2009) ICH S9: nonclinical evaluation for anticancer pharmaceuticals. http://www.ich.org
75. (2018) ICH S9 guideline: nonclinical evaluation for anticancer pharmaceuticals. questions and answers. http://www.ich.org
76. Ma A, Dun H, Song L, Hu Y, Zeng L, Bai J, Zhang G, Kinugasa F, Miyao Y, Sakuma S, Okimura K, Kasai N, Daloze P, Chen H (2014) Pharmacokinetics and pharmacodynamics of ASKP1240, a fully human anti-CD40 antibody, in normal and renal transplanted Cynomolgus monkeys. Transplantation 97(4):397–404. https://doi.org/10.1097/01.TP.0000440951.29757.bd
77. Mao CP, Brovarney MR, Dabbagh K, Birnbock HF, Richter WF, Del Nagro CJ (2013) Subcutaneous versus intravenous administration of rituximab: pharmacokinetics, CD20 target coverage and B-cell depletion in cynomolgus monkeys. PLoS One 8(11):e80533. https://doi.org/10.1371/journal.pone.0080533
78. Wyant T, Lackey A, Green M (2008) Validation of a flow cytometry based chemokine internalization assay for use in evaluating the pharmacodynamic response to a receptor antagonist. J Transl Med 6:76. https://doi.org/10.1186/1479-5876-6-76
79. Freeman DJ, McDorman K, Ogbagabriel S, Kozlosky C, Yang BB, Doshi S, Perez-Ruxio JJ, Fanslow W, Starnes C, Radinsky R (2012) Tumor penetration and epidermal growth factor receptor saturation by panitumumab correlate with antitumor activity in a preclinical model of human cancer. Mol Cancer 11:47. https://doi.org/10.1186/1476-4598-11-47
80. Lowe PJ, Hijazi Y, Luttringer O, Yin H, Sarangapani R, Howard D (2007) On the anticipation of the human dose in first-in-man trials from preclinical and prior clinical information in early drug development. Xenobiotica 37(10–11):1331–1354. https://doi.org/10.1080/00498250701648008
81. Stewart JJ, Green CL, Jones N, Liang M, Xu Y, Wilkins DE, Moulard M, Czechowska K, Lanham D, McCloskey TW, Ferbas J, van der Strate BW, Hogerkorp CM, Wyant T, Lackey A, Litwin V (2016) Role of receptor occupancy assays by flow cytometry in drug development. Cytometry B Clin Cytom 90(2):110–116. https://doi.org/10.1002/cyto.b.21355
82. Weber JS, Dummer R, de Pril V, Lebbe C, Hodi FS, Investigators MDX (2013) Patterns of onset and resolution of immune-related adverse events of special interest with ipilimumab: detailed safety analysis from a phase 3 trial in patients with advanced melanoma. Cancer 119(9):1675–1682. https://doi.org/10.1002/cncr.27969
83. Vermeire S, O'Byrne S, Keir M, Williams M, Lu TT, Mansfield JC, Lamb CA, Feagan BG, Panes J, Salas A, Baumgart DC, Schreiber S, Dotan I, Sandborn WJ, Tew GW, Luca D, Tang MT, Diehl L, Eastham-Anderson J, De Hertogh G, Perrier C, Egen JG, Kirby JA, van Assche G, Rutgeerts P (2014) Etrolizumab as induction therapy for ulcerative colitis: a randomised, controlled, phase 2 trial. Lancet 384 (9940):309–318. https://doi.org/10.1016/S0140-6736(14)60661-9
84. Martin DA, Churchill M, Flores-Suarez L, Cardiel MH, Wallace D, Martin R, Phillips K, Kaine JL, Dong H, Salinger D, Stevens E, Russell CB, Chung JB (2013) A phase Ib multiple ascending dose study evaluating safety, pharmacokinetics, and early clinical response of brodalumab, a human anti-IL-17R antibody, in methotrexate-resistant rheumatoid arthritis. Arthritis Res Ther 15(5):R164. https://doi.org/10.1186/ar4347
85. Vey N, Bourhis JH, Boissel N, Bordessoule D, Prebet T, Charbonnier A, Etienne A, Andre P, Romagne F, Benson D, Dombret H, Olive D (2012) A phase 1 trial of the anti-inhibitory KIR mAb IPH2101 for AML in complete remission. Blood 120(22):4317–4323. https://doi.org/10.1182/blood-2012-06-437558
86. Reilly M, Miller RM, Thomson MH, Patris V, Ryle P, McLoughlin L, Mutch P, Gilboy P, Miller C, Broekema M, Keogh B, McCormack W, van de Wetering de Rooij J (2013) Randomized, double-blind, placebo-controlled, dose-escalating phase I, healthy subjects study of intravenous OPN-305, a humanized anti-TLR2 antibody. Clin Pharmacol Ther 94(5):593–600. https://doi.org/10.1038/clpt.2013.150
87. Wei X, Gibiansky L, Wang Y, Fuh F, Erickson R, O'Byrne S, Tang MT (2017)

Pharmacokinetic and pharmacodynamic modeling of serum etrolizumab and circulating beta7 receptor occupancy in patients with ulcerative colitis. J Clin Pharmacol. https://doi.org/10.1002/jcph.1031

88. Wyant T, Estevam J, Yang L, Rosario M (2016) Development and validation of receptor occupancy pharmacodynamic assays used in the clinical development of the monoclonal antibody vedolizumab. Cytometry B Clin Cytom 90(2):168–176. https://doi.org/10.1002/cyto.b.21236

89. Sternebring O, Alifrangis L, Christensen TF, Ji H, Hegelund AC, Hogerkorp CM (2016) A weighted method for estimation of receptor occupancy for pharmacodynamic measurements in drug development. Cytometry B Clin Cytom 90(2):220–229. https://doi.org/10.1002/cyto.b.21277

90. Liang M, Schwickart M, Schneider AK, Vainshtein I, Del Nagro C, Standifer N, Roskos LK (2016) Receptor occupancy assessment by flow cytometry as a pharmacodynamic biomarker in biopharmaceutical development. Cytometry B Clin Cytom 90(2):117–127. https://doi.org/10.1002/cyto.b.21259

91. Goins CL, Chappell CP, Shashidharamurthy R, Selvaraj P, Jacob J (2010) Immune complex-mediated enhancement of secondary antibody responses. J Immunol 184(11):6293–6298. https://doi.org/10.4049/jimmunol.0902530

92. Bartelds GM, Krieckaert CL, Nurmohamed MT, van Schouwenburg PA, Lems WF, Twisk JW, Dijkmans BA, Aarden L, Wolbink GJ (2011) Development of antidrug antibodies against adalimumab and association with disease activity and treatment failure during long-term follow-up. JAMA 305(14):1460–1468. https://doi.org/10.1001/jama.2011.406

93. C. M (2014) The relevance of immunogenicity in preclinical development. J Bioanal Biomed 6:1):1–1):4

94. De Groot AS, McMurry J, Moise L (2008) Prediction of immunogenicity: in silico paradigms, ex vivo and in vivo correlates. Curr Opin Pharmacol 8(5):620–626. https://doi.org/10.1016/j.coph.2008.08.002

95. Krishna M, Nadler SG (2016) Immunogenicity to biotherapeutics – the role of anti-drug immune complexes. Front Immunol 7:21. https://doi.org/10.3389/fimmu.2016.00021

96. Singh SK (2011) Impact of product-related factors on immunogenicity of biotherapeutics. J Pharm Sci 100(2):354–387. https://doi.org/10.1002/jps.22276

97. Hwang WY, Foote J (2005) Immunogenicity of engineered antibodies. Methods 36(1):3–10. https://doi.org/10.1016/j.ymeth.2005.01.001

98. 1 ECBR (18 May 2017) Guideline on Immunogenicity assessment of therapeutic proteins. https://www.ema.europa.eu

99. FaDA (FDA) (Jan 2019) Immunogenicity testing of therapeutic protein products – developing and validating assays for anti-drug antibody detection. https://www.fda.gov

100. Shankar G, Arkin S, Cocea L, Devanarayan V, Kirshner S, Kromminga A, Quarmby V, Richards S, Schneider CK, Subramanyam M, Swanson S, Verthelyi D, Yim S, American Association of Pharmaceutical S (2014) Assessment and reporting of the clinical immunogenicity of therapeutic proteins and peptides-harmonized terminology and tactical recommendations. AAPS J 16(4):658–673. https://doi.org/10.1208/s12248-014-9599-2

101. Wadhwa M, Knezevic I, Kang HN, Thorpe R (2015) Immunogenicity assessment of biotherapeutic products: an overview of assays and their utility. Biologicals 43(5):298–306. https://doi.org/10.1016/j.biologicals.2015.06.004

102. (2005) ICH S8: immunotoxicity studies for human pharmaceuticals. http://www.ich.org

103. Peachee VL, Smith MJ, Beck MJ, Stump DG, White KL Jr (2014) Characterization of the T-dependent antibody response (TDAR) to keyhole limpet hemocyanin (KLH) in the Gottingen minipig. J Immunotoxicol 11(4):376–382. https://doi.org/10.3109/1547691X.2013.853716

104. Plitnick LM, Herzyk DJ (2010) The T-dependent antibody response to keyhole limpet hemocyanin in rodents. Methods Mol Biol 598:159–171. https://doi.org/10.1007/978-1-60761-401-2_11

105. Brennan FR, Morton LD, Spindeldreher S, Kiessling A, Allenspach R, Hey A, Muller PY, Frings W, Sims J (2010) Safety and immunotoxicity assessment of immunomodulatory monoclonal antibodies. MAbs 2(3):233–255

106. Galbiati V, Mitjans M, Corsini E (2010) Present and future of in vitro immunotoxicology in drug development. J Immunotoxicol 7(4):255–267. https://doi.org/10.3109/1547691X.2010.509848

107. Herzyk DJ, Haggerty HG (2018) Cancer immunotherapy: factors important for the evaluation of safety in nonclinical studies. AAPS J 20(2):28. https://doi.org/10.1208/s12248-017-0184-3
108. Lebrec H, Molinier B, Boverhof D, Collinge M, Freebern W, Henson K, Mytych DT, Ochs HD, Wange R, Yang Y, Zhou L, Arrington J, Christin-Piche MS, Shenton J (2014) The T-cell-dependent antibody response assay in nonclinical studies of pharmaceuticals and chemicals: study design, data analysis, interpretation. Regul Toxicol Pharmacol 69(1):7–21. https://doi.org/10.1016/j.yrtph.2014.02.008
109. Koup RA, Douek DC (2011) Vaccine design for CD8 T lymphocyte responses. Cold Spring Harb Perspect Med 1(1):a007252. https://doi.org/10.1101/cshperspect.a007252
110. Park H, Adamson L, Ha T, Mullen K, Hagen SI, Nogueron A, Sylwester AW, Axthelm MK, Legasse A, Piatak M Jr, Lifson JD, McElrath JM, Picker LJ, Seder RA (2013) Polyinosinic-polycytidylic acid is the most effective TLR adjuvant for SIV Gag protein-induced T cell responses in nonhuman primates. J Immunol 190(8):4103–4115. https://doi.org/10.4049/jimmunol.1202958
111. Loffredo J, Vuyyuru R, Spires V, Beyer S, Fox M, Ehrmann J, Taylor K, Engelhardt J, Korman A, Graziano R (2017) Non-fucosylated anti-CTLA-4 antibody enhances vaccine-induced t cell responses in a non-human primate pharmacodynamic vaccine model. J Immunother Cancer 5(2):P55
112. Leong ML, Newell EW (2015) Multiplexed peptide-MHC tetramer staining with mass cytometry. Methods Mol Biol 1346:115–131. https://doi.org/10.1007/978-1-4939-2987-0_9
113. Sims S, Willberg C, Klenerman P (2010) MHC-peptide tetramers for the analysis of antigen-specific T cells. Expert Rev Vaccines 9(7):765–774. https://doi.org/10.1586/erv.10.66
114. Pareja E, Tobes R, Martin J, Nieto A (1997) The tetramer model: a new view of class II MHC molecules in antigenic presentation to T cells. Tissue Antigens 50(5):421–428
115. Rodenko B, Toebes M, Hadrup SR, van Esch WJ, Molenaar AM, Schumacher TN, Ovaa H (2006) Generation of peptide-MHC class I complexes through UV-mediated ligand exchange. Nat Protoc 1(3):1120–1132. https://doi.org/10.1038/nprot.2006.121
116. Slota M, Lim JB, Dang Y, Disis ML (2011) ELISpot for measuring human immune responses to vaccines. Expert Rev Vaccines 10(3):299–306. https://doi.org/10.1586/erv.10.169
117. Svitek N, Taracha EL, Saya R, Awino E, Nene V, Steinaa L (2016) Analysis of the cellular immune responses to vaccines. Methods Mol Biol 1349:247–262. https://doi.org/10.1007/978-1-4939-3008-1_16
118. Disis ML, Wallace DR, Gooley TA, Dang Y, Slota M, Lu H, Coveler AL, Childs JS, Higgins DM, Fintak PA, dela Rosa C, Tietje K, Link J, Waisman J, Salazar LG (2009) Concurrent trastuzumab and HER2/neu-specific vaccination in patients with metastatic breast cancer. J Clin Oncol 27(28):4685–4692. https://doi.org/10.1200/JCO.2008.20.6789
119. Gulley JL, Arlen PM, Madan RA, Tsang KY, Pazdur MP, Skarupa L, Jones JL, Poole DJ, Higgins JP, Hodge JW, Cereda V, Vergati M, Steinberg SM, Halabi S, Jones E, Chen C, Parnes H, Wright JJ, Dahut WL, Schlom J (2010) Immunologic and prognostic factors associated with overall survival employing a poxviral-based PSA vaccine in metastatic castrate-resistant prostate cancer. Cancer Immunol Immunother 59(5):663–674. https://doi.org/10.1007/s00262-009-0782-8
120. Kirkwood JM, Lee S, Moschos SJ, Albertini MR, Michalak JC, Sander C, Whiteside T, Butterfield LH, Weiner L (2009) Immunogenicity and antitumor effects of vaccination with peptide vaccine+/−granulocyte-monocyte colony-stimulating factor and/or IFN-alpha2b in advanced metastatic melanoma: Eastern Cooperative Oncology Group Phase II Trial E1696. Clin Cancer Res 15(4):1443–1451. https://doi.org/10.1158/1078-0432.CCR-08-1231
121. Gray CM, Mlotshwa M, Riou C, Mathebula T, de Assis Rosa D, Mashishi T, Seoighe C, Ngandu N, van Loggerenberg F, Morris L, Mlisana K, Williamson C, Karim SA, Team CAIS (2009) Human immunodeficiency virus-specific gamma interferon enzyme-linked immunospot assay responses targeting specific regions of the proteome during primary subtype C infection are poor predictors of the course of viremia and set point. J Virol 83(1):470–478. https://doi.org/10.1128/JVI.01678-08
122. Tario JD Jr, Muirhead KA, Pan D, Munson ME, Wallace PK (2011) Tracking immune cell proliferation and cytotoxic potential using flow cytometry. Methods Mol Biol 699:119–164. https://doi.org/10.1007/978-1-61737-950-5_7

123. Wang X, Lebrec H (2017) Immunophenotyping: application to safety assessment. Toxicol Pathol 45(7):1004–1011. https://doi.org/10.1177/0192623317736742
124. Burchiel SW, Kerkvliet NL, Gerberick GF, Lawrence DA, Ladics GS (1997) Assessment of immunotoxicity by multiparameter flow cytometry. Fundam Appl Toxicol 38(1):38–54
125. Corsini E, House RV (2018) Evaluating cytokines in immunotoxicity testing. Methods Mol Biol 1803:297–314. https://doi.org/10.1007/978-1-4939-8549-4_18
126. Jackman JA, Meszaros T, Fulop T, Urbanics R, Szebeni J, Cho NJ (2016) Comparison of complement activation-related pseudoallergy in miniature and domestic pigs: foundation of a validatable immune toxicity model. Nanomedicine 12 (4):933–943. https://doi.org/10.1016/j.nano.2015.12.377
127. Szebeni J (2005) Complement activation-related pseudoallergy: a new class of drug-induced acute immune toxicity. Toxicology 216(2–3):106–121. https://doi.org/10.1016/j.tox.2005.07.023
128. Kawabata TT, Evans EW (2012) Development of immunotoxicity testing strategies for immunomodulatory drugs. Toxicol Pathol 40(2):288–293. https://doi.org/10.1177/0192623311430238
129. Hamczyk MR, Villa-Bellosta R, Andres V (2015) In vitro macrophage phagocytosis assay. Methods Mol Biol 1339:235–246. https://doi.org/10.1007/978-1-4939-2929-0_16
130. Lankveld DP, Van Loveren H, Baken KA, Vandebriel RJ (2010) In vitro testing for direct immunotoxicity: state of the art. Methods Mol Biol 598:401–423. https://doi.org/10.1007/978-1-60761-401-2_26
131. Narayanan PK, Li N (2019) In vitro monocyte/macrophage phagocytosis assay for the prediction of drug-induced thrombocytopenia. Curr Protoc Toxicol 79(1):e68. https://doi.org/10.1002/cptx.68
132. Jang YY, Cho D, Kim SK, Shin DJ, Park MH, Lee JJ, Shin MG, Shin JH, Suh SP, Ryang DW (2012) An improved flow cytometry-based natural killer cytotoxicity assay involving calcein AM staining of effector cells. Ann Clin Lab Sci 42(1):42–49

Chapter 13

The Quest for the Next-Generation of Tumor Targets: Discovery and Prioritization in the Genomics Era

Leonardo Mirandola, Franco Marincola, Gianluca Rotino, Jose A. Figueroa, Fabio Grizzi, Robert Bresalier, and Maurizio Chiriva-Internati

Abstract

Cancer immunotherapy has recently delivered impressive results. However, major roadblocks prevent the deployment of immune-oncology approaches to the majority of cancer patients, especially those with solid tumors. Chimeric antigen receptor (CAR) T-cell therapies, for example, are far from being successfully implemented.

A bottleneck in modern immunotherapy of cancer is the identification of molecules that allow effective targeting of the tumor while leaving normal tissues untouched. Historically, all major clinical developments have been focusing on tumor-associated antigens (TAAs). The landscape of TAAs, however, is restricted, with TAAs limited to cancer/testis antigens, which showed promising pre-clinical and clinical results but are targetable in only a limited number of malignancies. Furthermore, the magnitude of the immune response against hyper-expressed antigens is hindered by the general low affinity of T-cell receptors (TCRs) for self-antigens.

Recent efforts have, therefore, been directed towards the discovery and validation of neoantigens, that arise from mutations in cancer cells. Emerging alternatives to neoantigens are mutational hotspot antigens and cancer-specific isoforms of widespread proteins. Here, we describe the advantages and the drawbacks of each antigenic class in the field of immunotherapy, as well as novel in silico prediction tools for the high-throughput analysis of large "omics" datasets, which will dramatically speed up the discovery of new cancer-specific immunodominant antigens with high potential of being translated to the clinic.

Key words Cancer antigens, Immunotherapy, Neoantigens, Splice variants, T-cell response, In silico prediction

1 Introduction

Immunotherapy has shown impressive potential for the treatment of hematologic and solid malignancies. The best known and most evident benefits have been so far obtained with immune checkpoint inhibitors (ICI) and CAR T-cells. The 2018 Noble Prize in Medicine and Physiology for work in this field has drawn the general public's attention to the importance of the role played by the immune system in the prognosis and treatment outcome of cancer.

In order to direct the immune system against the tumor and not the host, we need to discover tumor-specific biomarkers, so that the immune system can distinguish malignant from normal cells. This discovery is complex, and the amount of available "tags" is limited and thus the need for novel methods to find them.

The discovery and prioritization of putative targets of anti-tumoral cell-mediated responses have been minimal compared with the parallel field of infectious disease. The enthusiasm about the utilization of CAR T-cells is growing due to the observation of unprecedented response rates in patients receiving CD19-directed CAR-T for both acute lymphoid leukemia (AML) and non-Hodgkin lymphoma (NHL) [1, 2]. It is likely that we will soon see additional approvals by the Food and Drug Administration for CD19 CAR T-cells and for B-cell maturation antigen (BCMA) CAR T-cells in other indications such as multiple myeloma, given the promising early trial results [3]. However, many investigators are concerned that the current enthusiasm for CD19 and BCMA may fade if the discovery and validation of additional targets for these and other indications do not quickly follow. The ability to further extend these approaches to a broad range of alternative targets for cancers of different histological origins, especially solid tumors, is pivotal for the advancement of cancer immunotherapy. Therefore, applying novel systems to identify and validate cancer-specific antigens will yield a greater pipeline of targets for cancer-specific T-cells and vaccines.

In this chapter, we will discuss the different classes of cancer targetable antigens, namely TAAs, neoantigens (Fig. 1), alternative splicing-derived antigens, and present innovative methods of "omics" data mining integrating different datasets, as well as the use of innovative neural networks to predict the immune-dominant antigens for optimal anti-tumoral T-cell response.

2 Tumor-Associated Antigens

Tumor-associated antigens (TAAs) are proteins or glycoproteins expressed almost exclusively by tumor cells. TAAs can be found in any subcellular compartment: membrane, cytoplasm, nucleus, or they can be secreted by the tumor cells. The TAAs that are of interest for immunotherapy are those that are differentially expressed in tumors compared to corresponding normal tissues. TAAs are generally divided in the following categories:

- *Oncofetal*—these are typically only expressed in fetal tissues and in neoplastic somatic cells.
- *Overexpressed*—expressed by both normal and neoplastic cells, but with a higher level of expression in neoplasia compared with non-cancerous tissues.

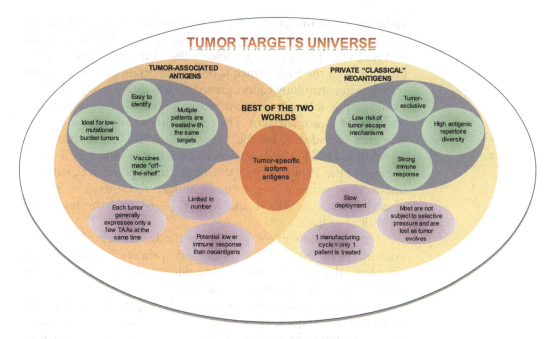

Fig. 1 The universe of cancer targets. Tumor-associated antigens are advantageous because they are shared by many patients and generally tend to be easy to identify (no need of next-generation sequencing and complex in silico analyses). Additionally, they tend to be amenable to off the shelf manufacturing because the same peptide sequences can be used for all TAA+ patients with the same HLA allele. However, TAAs generally tend to be expressed and presented by a fraction of the tumor mass (close but not equal to 100%), they are limited in number, and they afford for lower immunological responses compared with mutated antigens or microbial antigens, because they are self-proteins and central and/or peripheral tolerance may hinder the adaptive immunity against them. Private neoantigens afford for potent immunological response due to the lack of tolerance against them because the immune system sees them as non-self. In addition, because they are exclusive to the tumor, the chances of on-target/off-tumor side effects are dramatically reduced. They are potentially unlimited in number, making possible to target almost any patient. On the other hand, they require a complex and expensive patient-specific analysis of tumor and normal genome and tumor transcriptome, and they are not amenable to off the shelf manufacturing. Kiromic is focusing its next-generation anticancer immunotherapies on the artificial intelligence-driven prediction of a third class of tumor antigens, which comprises that best of the two worlds just described. Shared tumor-specific splice isoform antigens afford for the strong immunological response and high patient population coverage typical of private and shared neoantigens but also combine the off the shelf manufacturing capability of the tumor-associated antigens

- *Differentiation*—expressed on tumor cells, and during at least some stage of differentiation on nonmalignant cells of the cell lineage from which the tumor developed.
- *Cancer-testis*—expressed only by cancer cells and adult reproductive tissues such as testes and placenta.

Historically, TAAs were the first to be identified and validated in cancer immunotherapy. Accordingly, the main obstacle tumor immunologists had to overcome at the time was to study and reprogram the immune system–tumor cell interactions without a prioritized list of targets to model the tumor-immune system interactions. In the late 1980s/early 1990s, a great deal of experiments was performed in murine models with foreign antigen (OVA, SV40-T)-expressing tumor cells. Although these experiments proved the basic principle of the importance of the immune response in controlling cancer growth, it was difficult to predict if it would be possible to translate those findings into immunity against self-derived tumor antigens. The equivalent of the periodic table of elements introduced by the Russian chemist Dmitri Ivanovich Mendeleev (1834–1907) was needed for tumor antigens. Toward the end of the 1980s, such a table began to be populated, with the discovery of putative human tumor antigen recognized by T lymphocytes, such as MUC1 and MAGE [4–6], and the list is still expanding to date.

The development of genetic methods, combined with the availability of tumor cell lines and primary tumor clones, afforded—the identification of molecules in cancer cells that were a source of T-cell-recognized antigens.

Rapidly expanding and continuously improving cellular and molecular techniques, developed in other disciplines and transferred to immunology, allowed researchers to discover and characterize hundreds of molecules that could be targeted by CD8+ CTL, CD4+ helper T-cells, and antibodies. These potential targets have been organized into the categories listed above. One of the most attractive categories is cancer-testis antigens (CTAs); this includes some of the best characterized and validated targets, such as of MAGE, GAGE, and BAGE (originally identified on melanomas but also expressed by many other cancers) and NY-ESO-1, together with novel, "structural protein" CTAs: SP17, AKAP4, PTTG1, Ropporin-1, Span-XB1 [7–17]. Differentiation antigens are another large family of molecules including CEA, gp100, MART-1, PSA, and tyrosinase. Overexpressed antigens such as HER-2/neu, TERT, MUC1, mesothelin, PSA, PSMA, survivin, WT-1, p53, and cyclin B1 represent, however, the broadest category.

Apart from "oncofetal antigens", which are present only in a minority of tumors, the value of TAAs in anti-cancer immunotherapy relies solely on their preferential expression by transformed cells. Their basal expression in normal tissues dampens the immune response against these proteins due to central and peripheral tolerance mechanisms, with only low-avidity T-cells able to survive. On the other hand, the use of high-avidity transgenic TCR T-cells may cause severe autoimmune reactions [18]. Compared with other classes of TAAs, CTAs have attracted more attention thanks to their wide expression in different tumors and their restriction to tumor cells. Moreover, the expression of CTAs has been shown to be correlated with intra-tumoral T-cell infiltration and improved prognosis of cancer patients though these lymphocytes might be functionally impaired. It has also been established that adoptive transfer of lymphocytes genetically engineered with an NY-ESO-1-directed TCR is able to induce tumor regression.

Several HLA class I- and class II-restricted peptides have been identified from each of these antigens, and most of them are also B-cell epitopes, recognized by patients' circulating auto-antibodies. However, despite the potentially broad application of immunotherapies against TAA self-antigens, the expansion of T-cells specific for them has proven difficult, because of central and peripheral tolerance which have been associated with poor clinical benefits, and the accompanying risk of developing autoimmune disorders [18, 19].

Tumor-specific antigens (TSAs) have recently gained the attention of researchers and clinicians compared to TAAs, given their great potential for overcoming the obstacles associated with the TAAs-targeted approach described above. TSAs are absent from the normal genome and therefore from normal cells and include antigens derived from viral oncogenic proteins [oncoviral proteins including Simian virus 40 (SV40), or E6/E7 from human papilloma virus (HPV)], or from proteins that are encoded by genes whose sequences are altered by somatic mutations or DNA rearrangements. Since virus-derived proteins are only found in tumors caused by oncogenic viruses (viruses suspected of causing cancer in animals and humans), their usefulness is limited. However, tumors are generally prone to acquire mutations during carcinogenesis and progression that spontaneously produce large numbers of potentially antigenic determinants identifiable efficiently with modern technologies. They have, therefore, been at the center of attention for the development of next-generation cancer immunotherapy products.

3 "Passenger" Neoantigens

Tumor somatic mutations that are directly linked to or cause the transformation process are known as driver mutations, while those that occur secondarily to the genetic instability of rapidly expanding tumor cells are defined as "passenger" mutations [20]. Such muted antigens have great potential for immunotherapy because these neoantigens and neoepitopes are not "protected" by thymic selection and central tolerance.

Passenger neoantigens are randomly produced by sporadic mutations as a consequence of errors of the DNA repair machinery, and consequently their selection is predominantly patient-specific also defined as "private".

As a consequence of their private nature, potential neoantigen-derived peptides (i.e., neoepitopes) can only be discovered by analyzing each individual patient's tumor mutanome by whole exome sequencing (WES) and next-generation RNA sequencing

(RNA-seq) [21, 22]. Depending on the tumor type, the number of potential private neoantigens can vary from only a few (i.e., astrocytoma) to several thousand (melanomas and lung cancers) [23].

Unfortunately, there is no consensus regarding how to prioritize neoantigens, and the lack of effective prediction models makes it difficult to predict clinical benefits. Today, a wide array of software and in silico tools are available to predict immunogenic neoantigens, but each has its own strengths and weaknesses, and it has been difficult to identify which tools or which combinations of tools works best [24].

Despite the difficulties, vaccination against tumor-specific neoantigens minimizes the potential for induction of central and peripheral tolerance as well as the risk of autoimmunity. Neoantigen-based cancer vaccines have recently showed marked therapeutic potential in both preclinical and early-phase clinical studies.

Compared to non-mutated self-antigens, neoantigens could be recognized as non-self by the patient's immune system, and they have therefore potentially increased specificity, efficacy, and safety [25]. The immunogenicity of neoantigens in terms of their ability to elicit a productive T-cell response has long been seen in human trials [26], as neoantigen-specific CTLs are the most powerful tumor-rejection T-cells [27–30]. However, naturally occurring neoantigen-specific CTLs are rare, likely because of low clonal frequency and suboptimal presentation of neoantigens by antigen-presenting cells (APCs) [23, 31]. A recent study in advanced melanoma using a poly-neoepitope-encoding mRNA vaccine showed that this neoantigen-based vaccination approach is safe and effective [32], with vaccine-induced, potent T-cell responses against multiple neoantigens detected in all patients, without toxicities.

In summary, the main drawback of private neoantigen-directed approaches is the limited number of neoantigens that can be effectively presented by tumor cells to elicit a substantive T-cell response (recent estimates suggest around 1%).

Moreover, given the size of the human exome (around 30 Mbs), the chances that a somatic coding mutation will occur in more than one patient is very small.

Yet, not all somatic mutations occur "randomly", and some are shared by multiple patients with a given cancer type: this is the case of driver neoantigens, also known as "shared" neoantigens.

4 "Driver" Neoantigens

Mutations that affect the function of a protein to promote oncogenesis, drug resistance, and tumor survival, are collectively referred to as "driver mutations". They can be consistently detected in several patients with the same cancer type, and they typically occur in "hotspot" sequences.

Driver mutations typically involve a single or a few amino-acid substitutions, and they tend to be conserved at metastatic sites [33]. The safety and effectiveness of cancer immunotherapies targeting shared neoantigens has been proven in the clinical setting. A Phase II study in patients with stage I–III lung adenocarcinoma showed that a vaccine targeting K-ras hotspot mutations was well tolerated and immunogenic as consolidation therapy. In another study in lung, colon, breast, ovarian, head and neck, pancreatic, esophageal, and gastric cancer, an immunization with a peptide of either p53 or K-ras surrounding the tumor-specific hotspot mutations (NCI trial T93-0148) indicated that the mutated peptide vaccination was feasible without any toxicity. Cytotoxic T lymphocytes (CTLs) and cytokine responses specific to a given mutation could be induced, and cellular immunity to mutant p53 and K-ras was associated with longer survival [34].

An advantage of immunotherapies targeting neoepitopes originating from driver mutations is that the immune responses they induce are likely more effective compared with those against private neoantigens because proteins produced by genes affected by driver mutations are most likely essential for tumor survival and are, therefore, homogenously expressed by the whole tumor cell population. An example of neoepitopes derived from well-known mutations in well-known oncogenes is KRAS mutated at codon 12 or mutated p53 [35, 36]. Another oncogenic alteration that frequently occurs in melanoma is the V599E in the BRAF protein, which originates a mutated epitope that can be recognized by T-cells [37, 38]. Furthermore, in some hematological tumors, mutations in JAK2 (V617F) in calreticulin can give rise to spontaneous T-cell responses [39, 40]. These, along with the histone H3 K27M mutation in glioma [41, 42], could be interesting targets of newly designed immunotherapies for such diseases. Furthermore, in a humanized murine model of glioma, it has been shown that a vaccine targeting the isocitrate dehydrogenase-1 (IDH1) R132H mutation promoted specific T-cell responses and controlled tumor growth [43].

It should be noted that mutations and neoepitopes are related but they are distinct, with most driver mutations not resulting in neoantigens. The latter is the result of a mutated protein presentation process, which largely depends on the availability of the correct HLA allele. Therefore, despite hotspot mutations that are a rather frequent genetic event in cancers, the number of shared neoepitopes is dramatically reduced by the enormous HLA diversity. A recent genomic analysis of more than 63,220 cancer cases including multiple tumor types predicted that the neoepitopes that could be shared among the most common HLA-A and HLA-B alleles are still rare, albeit more common than those originating from passenger mutations [44].

Shared neoepitopes, however, are not fully private, and represent an important technical advantage for vaccine manufacturing and potentially for achieving a long-term response to prevent the rise of escape mutants which is likely the case with respect to passenger mutations, because the latter do not provide the cancer cells with any survival advantage.

5 Data-Mining Tools for In Silico Prioritization of Shared Cancer Antigens

The availability of RNA-Seq in recent years has empowered immunologists and cancer biologists with a powerful method for transcriptomic analysis [45]. RNA-Seq provides data that is applicable to many different fields of research, such as understanding gene functions, defining biological patterns, finding drug targets and discovering biomarkers for disease classification, diagnosis, and molecular targeting. Building on such breakthrough and high-throughput technology, the cancer genome atlas (TCGA) [46] and genotype-tissue expression (GTEx) [47, 48] collected and categorized RNA-Seq data for several thousands of cancer and non-cancer samples, providing an unprecedented dataset visualization and analyses tools.

The TCGA contains RNA-Seq data for almost 10,000 tumor samples spanning 33 cancer types and 726 adjacent normal tissues samples. Such an imbalance between the tumor and the non-tumoral counterpart number of samples can cause biases in differential analyses. This imbalance could be solved by computational methods which compare the TCGA and the GTEx data points. Indeed, the GTEx project generated RNA-Seq data for over 8000 non-tumoral tissues. To this end, i.e., to make data from different sources and pipelines more compatible, the University of California Santa Cruz (UCSC) Xena project has re-calculated all raw data from TCGA and GTEx based on a standard pipeline, thus allowing for reducing the bias in comparing samples and cohorts from the two datasets.

Gene expression differential analysis is a supervised analysis method, aimed at finding tumor-specific genes by comparing tumor to normal groups. Accurate and specific differential expression analysis is key to any in silico prediction tools for the identification and prioritization of novel tumor antigens.

Currently, Xena, cBioPortal, HPA, and Expression Atlas are the main visualization and analysis tools for gene expression analysis. These web servers are widely used and are the gold standard for the analysis of TCGA and GTEx data, and yet many expression analysis functions are not adequately addressed by these tools. Differential expression analyses are commonly requested by immune-

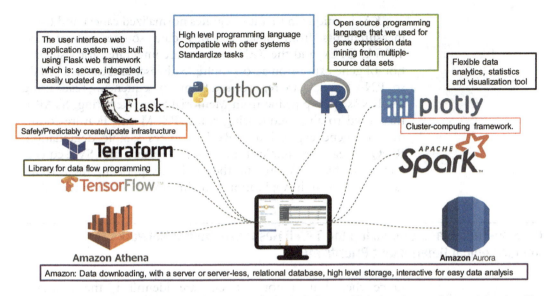

Fig. 2 KAI's ecosystem. KAI is a cloud-based ecosystem including multiple functions and large datasets with an unprecedented integration efficiency. This figure summarizes the multi-platform nature of KAI, with can seamlessly work the researcher through the whole antigen discovery and prioritization process, from data mining and meta-analysis to fine-tuning of the immune response by predicting the most immunogenic peptides to maximize the adaptive anti-tumoral response in a powerful yet safe manner

oncologists, however, this function is not available in cBioPortal, Xena or HPA, and the Expression Atlas does not allow detailed tumor-normal comparisons.

Kiromic Biopharma (http://kiromic.com/) has recently developed an innovative and comprehensive artificial intelligence tool, KAI, a cognitive machine and deep learning platform that extracts information from digital libraries consisting of clinical studies, and genomic and proteomic data, that is sourced from online databases (The Human Protein ATLAS, TCGA, GEO) and the Company's proprietary datasets (Fig. 2). It addresses one of the main challenges in today's clinical pipeline development which is disease-restricted target identification. Through KAI, genes can be identified that are highly and specifically expressed in the tumor of interest while providing its distribution and methylation status across the entire patient population. It also maps out, by virtue of a novel capsule artificial neural network, the exact portion of the gene that will elicit an immune response and hence can be used for the generation of antibodies, vaccines, T-cell therapies, and diagnostics for diseases such as cancer, metabolic syndromes, infectious disease, and autoimmune disorders. The software performs meta- and convolution-analyses while standardizing and normalizing data across multiple and variable experimental platforms such as RNA-seq and microarrays in normal vs. disease tissues, which allows for the visualization of consistent and accurate results in a user-friendly fashion.

KAI is unique in that it integrates normalized cancer and tissue measurements from heterogeneous data sources. Data from TCGA, GTEx, and the XENA project are integrated for visualization of gene expression across multiple diseases and cancers. To do so, RNA-Seq data are reprocessed from raw sequencing reads using the RNAseqCB pipeline from Memorial Sloan Kettering. STAR is then used to align sequencing reads, RSEM and FeatureCounts quantify gene expression, mRIN evaluates sample degradation, RSeQC measures sample strandness and quality, and SVAseq corrects batch biases. At present, the database supporting the Kiromic Genomic Research Application contains over 1.5 billion records.

6 Fine-Tuning and Optimizing the T-Cell Response: Neural Networks for Immunological Determinant Prioritization

Once the ideal tumor antigens are identified, the immune-dominant epitopes for any given HLA set have to be identified. This is done using common online tools such as those from IEDB [49], namely NetCTL-Pan, NetCTL-Chop, NetCTL, and NetCTLIIPan, EPIMHC [50], Antijen [51]. All these are data-driven methods for peptide-MHC binding prediction, and they are based on peptide sequences that are known to bind to MHC molecules from experimental "wet lab" results. In general, the non-linearity of peptide-MHC binding data [52] makes machine learning a necessary tool for accurate predictions. It is known that machine learning systems are trained with a list of peptides that either bind or do not bind to a given HLA, and the most relevant examples of machine learning-based peptide prediction models are artificial neural networks (ANNs). A further challenge is the fact that MHC binding affinity is only one of the factors shaping the peptidome of a given tumor, the other being the likelihood of proteasome processing and TAP binding. As a result, models that combine HLA affinity, proteasome processing scores and TAP binding scores increase the T-cell epitope predictive rate in comparison to just peptide-binding to MHC I [53].

At present, which one or set of algorithms should be used remains unclear. All algorithms have intrinsic limitations due to HLA alleles with insufficient binding data. In addition, the immunogenic potency of peptides does not solely depend on the peptide-MHC binding intensity or likelihood of processing by the antigen-presenting machinery [54], because the processed and presented peptide must be recognized by a specific TCR in order to generate a meaningful anti-cancer immune response. An in silico model, such as that being developed at Kiromic Biopharma, that is able to integrate the likelihood of peptide processing with that of TCR activation would solve the problem of high false-positive rates seen with current prediction methods, i.e., the high frequency with which currently available models predict peptides that are presented by the target cells but go unnoticed by T-cells.

7 The Future of Shared Cancer Targets: Isoantigens, Antigens Derived from Cancer-Specific Alternative Splicing Mechanisms

If a tumor-specific peptide could be identified that is presented by one of the patient's HLA alleles, an ideal "public" tumor-specific antigen shared across patients with that allele would be created. Such a peptide would solve the practical and regulatory obstacles currently seen with neoantigens, and it would be the "first-in-class" non-tumor associated antigen/non-neoantigen cancer-specific target allowing off-the-shelf manufacturing. Because of their private nature, this would not be possible with classic neoantigens and it is unlikely even with shared neoantigens.

Alternative splicing is an essential regulatory process of gene expression, and it affects most human genes [55]. Alternative splicing is a process by which exons of a gene may be included or excluded in the matured mRNAs, resulting in multiple distinct transcript forms, which in turn generate various protein isoforms [56]. A number of different mechanisms of alternative splicing have been described: use of alternative promoters, exon skipping, mutually exclusive exons, exon scrambling, alternative 5' and 3' splice sites, retained introns, and alternative polyadenylation [55, 56].

The majority of genes are alternatively spliced, and there are many isoforms specifically associated with cancer progression and metastasis. Splicing factors are de-regulated in cancer, and in some cases, they directly contribute to driving cancer progression. Cancer-specific formation of alternative transcripts has been proposed as a potential source of biomarkers for diagnosis [57, 58] and cancer stratification [59, 60]. It has also been hypothesized that tumor progression is associated with an altered splicing control, hence cancer-specific isoforms are likely to be expressed by the majority of cells in the tumor mass and to be conserved in metastases, unlike neoantigens.

Altered splicing patterns in cancers have been associated with splicing regulators [56, 61], such as RBFOX2, PTB/PTBP1, and SRSF1, which were found to cause splicing changes in multiple genes [57, 61].

The availability of whole transcriptome sequencing (RNA-seq, Iso-Seq) and the development of bioinformatics analysis tools to analyze and handle large sets of data have empowered the detection and the measurement of genes at the splicing level. Despite these advancements, a pipeline for the identification, prioritization, and validation of cancer-specific isoforms for immunotherapy is not currently commercially available.

In order to overcome the problem of identifying shared, "common" cancer-specific antigens derived from alternative splicing and cancer-specific isoforms formation, Kiromic Biopharma has also developed a fully integrated in silico prediction model of cancer-specific isoforms. This tool, CancerSplice, is part of KAI, the Company's artificial intelligence platform. CancerSplice allows for the prediction and prioritization of isoantigens which could serve as a novel source of tumor targets, highly specific for neoplastic cells but without the drawback of being highly patient-specific as the aforementioned neoantigens. CancerSplice allows the user to select a TCGA tissue type along with thresholds for filtering isoforms (minimum tumor TPM, maximum normal TPM). Based on the tissue selected, CancerSplice displays a sorted list of isoforms with the strongest tumor to normal differences. Differential analysis is performed and used to generate types of lists: a. isoforms expressed in tumors but not expressed in natural counterparts and b. isoforms expressed at much higher levels in tumors than natural counterparts. CancerSplice then allows the user to click on an isoform in the list to select a specific isoform to display in a detailed panel, which shows the multi-sequence alignment for the isoform and all the other isoforms of the gene. Finally, CancerSplice also shows a box plot by tissue of expression of the isoform in normal TCGA tissue and a box plot of the matching isoform in the GTEX normal data. The sequence of amino acids that is specific for the selected cancer isoforms is then directly fed to KAI's artificial neural capsule network for peptide design and prioritization across all major HLA-A, HLA-B, and HLA-DRB1 alleles.

References

1. Grupp SA et al (2013) Chimeric antigen receptor-modified T cells for acute lymphoid leukemia. N Engl J Med 368:1509–1518. https://doi.org/10.1056/NEJMoa1215134.
2. Neelapu SS et al (2017) Axicabtagene ciloleucel CAR T-cell therapy in refractory large B-cell lymphoma. N Engl J Med 377:2531–2544. https://doi.org/10.1056/NEJMoa1707447.
3. Carpenter RO et al (2013) B-cell maturation antigen is a promising target for adoptive T-cell therapy of multiple myeloma. Clin Cancer Res 19:2048–2060. https://doi.org/10.1158/1078-0432.Ccr-12-2422.
4. Gendler SJ et al (1987) Cloning of partial cDNA encoding differentiation and tumor-associated mucin glycoproteins expressed by human mammary epithelium. Proc Natl Acad Sci 84:6060–6064. https://doi.org/10.1073/pnas.84.17.6060
5. Siddiqui J et al (1988) Isolation and sequencing of a cDNA coding for the human DF3 breast carcinoma-associated antigen. Proc Natl Acad Sci 85:2320–2323. https://doi.org/10.1073/pnas.85.7.2320
6. van der Bruggen P et al (1991) A gene encoding an antigen recognized by cytolytic T lymphocytes on a human melanoma. Science 254:1643–1647. https://doi.org/10.1126/science.1840703
7. Mirandola L et al (2015) Novel antigens in non-small cell lung cancer: SP17, AKAP4, and PTTG1 are potential immunotherapeutic targets. Oncotarget 6:2812–2826. https://doi.org/10.18632/oncotarget.2802
8. Mirandola L et al (2017) Cancer testis antigen sperm protein 17 as a new target for triple negative breast cancer immunotherapy.

8. Oncotarget 8:74378–74390. https://doi.org/10.18632/oncotarget.20102
9. Mirandola L et al (2011) Tracking human multiple myeloma xenografts in NOD-Rag-1/IL-2 receptor gamma chain-null mice with the novel biomarker AKAP-4. BMC Cancer 11:394. https://doi.org/10.1186/1471-2407-11-394
10. Schutt C et al (2012) Immunological treatment options for locoregionally advanced head and neck squamous cell carcinoma. Int Rev Immunol 31:22–42. https://doi.org/10.3109/08830185.2011.637253
11. Chiriva-Internati M, Bot A (2015) A new era in cancer immunotherapy: discovering novel targets and reprogramming the immune system. Int Rev Immunol 34:101–103. https://doi.org/10.3109/08830185.2015.1015888
12. Chiriva-Internati M et al (2008) AKAP-4: a novel cancer testis antigen for multiple myeloma. Br J Haematol 140:465–468. https://doi.org/10.1111/j.1365-2141.2007.06940.x
13. Chiriva-Internati M, Wang Z, Pochopien S, Salati E, Lim SH (2003) Identification of a sperm protein 17 CTL epitope restricted by HLA-A1. Int J Cancer 107:863–865. https://doi.org/10.1002/ijc.11486
14. Chiriva-Internati M et al (2002) Sperm protein 17 (Sp17) is a suitable target for immunotherapy of multiple myeloma. Blood 100:961–965. https://doi.org/10.1182/blood-2002-02-0408
15. Chiriva-Internati M, Wang Z, Salati E, Timmins P, Lim SH (2002) Tumor vaccine for ovarian carcinoma targeting sperm protein 17. Cancer 94:2447–2453. https://doi.org/10.1002/cncr.10506
16. Chiriva-Internati M et al (2010) Cancer testis antigen vaccination affords long-term protection in a murine model of ovarian cancer. PLoS One 5:e10471. https://doi.org/10.1371/journal.pone.0010471
17. Grizzi F, Chiriva-Internati M (2013) Translating sperm protein 17 as a target for immunotherapy from the bench to the bedside in the light of cancer complexity. Tissue Antigens 81:116–118. https://doi.org/10.1111/tan.12052
18. Parkhurst MR et al (2011) T cells targeting carcinoembryonic antigen can mediate regression of metastatic colorectal cancer but induce severe transient colitis. Mol Ther 19:620–626. https://doi.org/10.1038/mt.2010.272
19. Hailemichael Y et al (2013) Persistent antigen at vaccination sites induces tumor-specific CD8 (+) T cell sequestration, dysfunction and deletion. Nat Med 19:465–472. https://doi.org/10.1038/nm.3105
20. Greenman C et al (2007) Patterns of somatic mutation in human cancer genomes. Nature 446:153–158. https://doi.org/10.1038/nature05610
21. Castle JC et al (2012) Exploiting the mutanome for tumor vaccination. Cancer Res 72:1081–1091. https://doi.org/10.1158/0008-5472.Can-11-3722
22. Karasaki T et al (2017) Prediction and prioritization of neoantigens: integration of RNA sequencing data with whole-exome sequencing. Cancer Sci 108:170–177. https://doi.org/10.1111/cas.13131
23. Alexandrov LB et al (2013) Signatures of mutational processes in human cancer. Nature 500:415–421. https://doi.org/10.1038/nature12477
24. (2017) The problem with neoantigen prediction. Nat Biotechnol 35:97. doi:https://doi.org/10.1038/nbt.3800
25. Yarchoan M, Johnson BA 3rd, Lutz ER, Laheru DA, Jaffee EM (2017) Targeting neoantigens to augment antitumour immunity. Nat Rev Cancer 17:209–222. https://doi.org/10.1038/nrc.2016.154
26. Wolfel T et al (1995) A p16INK4a-insensitive CDK4 mutant targeted by cytolytic T lymphocytes in a human melanoma. Science 269:1281–1284
27. Matsushita H et al (2012) Cancer exome analysis reveals a T-cell-dependent mechanism of cancer immunoediting. Nature 482:400–404. https://doi.org/10.1038/nature10755
28. Tran E et al (2014) Cancer immunotherapy based on mutation-specific CD4+ T cells in a patient with epithelial cancer. Science 344:641–645. https://doi.org/10.1126/science.1251102
29. Tran E et al (2016) T-cell transfer therapy targeting mutant KRAS in cancer. N Engl J Med 375:2255–2262. https://doi.org/10.1056/NEJMoa1609279
30. Tran E, Robbins PF, Rosenberg SA (2017) 'Final common pathway' of human cancer immunotherapy: targeting random somatic mutations. Nat Immunol 18:255–262. https://doi.org/10.1038/ni.3682
31. Zhu G, Zhang F, Ni Q, Niu G, Chen X (2017) Efficient nanovaccine delivery in cancer immunotherapy. ACS Nano 11:2387–2392. https://doi.org/10.1021/acsnano.7b00978
32. Sahin U et al (2017) Personalized RNA mutanome vaccines mobilize poly-specific therapeutic immunity against cancer. Nature 547:222–226. https://doi.org/10.1038/nature23003
33. Makohon-Moore AP et al (2017) Limited heterogeneity of known driver gene mutations

among the metastases of individual patients with pancreatic cancer. Nat Genet 49:358–366. https://doi.org/10.1038/ng.3764
34. Carbone DP et al (2005) Immunization with mutant p53- and K-ras-derived peptides in cancer patients: immune response and clinical outcome. J Clin Oncol 23:5099–5107. https://doi.org/10.1200/JCO.2005.03.158
35. Shono Y et al (2003) Specific T-cell immunity against K-ras peptides in patients with pancreatic and colorectal cancers. Br J Cancer 88:530–536. https://doi.org/10.1038/sj.bjc.6600697
36. Ichiki Y et al (2004) Simultaneous cellular and humoral immune response against mutated p53 in a patient with lung cancer. J Immunol 172:4844–4850
37. Sharkey MS, Lizee G, Gonzales MI, Patel S, Topalian SL (2004) CD4(+) T-cell recognition of mutated B-RAF in melanoma patients harboring the V599E mutation. Cancer Res 64:1595–1599
38. Andersen MH et al (2004) Immunogenicity of constitutively active V599EBRaf. Cancer Res 64:5456–5460. https://doi.org/10.1158/0008-5472.Can-04-0937
39. Holmstrom MO et al (2017) The JAK2V617F mutation is a target for specific T cells in the JAK2V617F-positive myeloproliferative neoplasms. Leukemia 31:495–498. https://doi.org/10.1038/leu.2016.290
40. Holmstrom MO et al (2018) The calreticulin (CALR) exon 9 mutations are promising targets for cancer immune therapy. Leukemia 32:429–437. https://doi.org/10.1038/leu.2017.214
41. Schwartzentruber J et al (2012) Driver mutations in histone H3.3 and chromatin remodelling genes in paediatric glioblastoma. Nature 482:226–231. https://doi.org/10.1038/nature10833
42. Bender S et al (2013) Reduced H3K27me3 and DNA hypomethylation are major drivers of gene expression in K27M mutant pediatric high-grade gliomas. Cancer Cell 24:660–672. https://doi.org/10.1016/j.ccr.2013.10.006
43. Schumacher T et al (2014) A vaccine targeting mutant IDH1 induces antitumour immunity. Nature 512:324–327. https://doi.org/10.1038/nature13387
44. Hartmaier RJ et al (2017) Genomic analysis of 63,220 tumors reveals insights into tumor uniqueness and targeted cancer immunotherapy strategies. Genome Med 9:16. https://doi.org/10.1186/s13073-017-0408-2

45. Wang Z, Gerstein M, Snyder M (2009) RNA-Seq: a revolutionary tool for transcriptomics. Nat Rev Genet 10:57–63. https://doi.org/10.1038/nrg2484
46. Weinstein JN et al (2013) The Cancer Genome Atlas Pan-Cancer analysis project. Nat Genet 45:1113–1120. https://doi.org/10.1038/ng.2764
47. genomics H (2015) The Genotype-Tissue Expression (GTEx) pilot analysis: multitissue gene regulation in humans. Science 348:648–660. https://doi.org/10.1126/science.1262110
48. GTEx Consortium (2013) The Genotype-Tissue Expression (GTEx) project. Nat Genet 45:580–585. https://doi.org/10.1038/ng.2653
49. Vita R et al (2015) The immune epitope database (IEDB) 3.0. Nucleic Acids Res 43:D405–D412. https://doi.org/10.1093/nar/gku938
50. Molero-Abraham M, Lafuente EM, Reche P (2014) Customized predictions of peptide-MHC binding and T-cell epitopes using EPIMHC. Methods Mol Biol (Clifton, NJ) 1184:319–332. https://doi.org/10.1007/978-1-4939-1115-8_18
51. Toseland CP et al (2005) AntiJen: a quantitative immunology database integrating functional, thermodynamic, kinetic, biophysical, and cellular data. Immun Res 1:4. https://doi.org/10.1186/1745-7580-1-4
52. Lafuente EM, Reche PA (2009) Prediction of MHC-peptide binding: a systematic and comprehensive overview. Curr Pharm Des 15:3209–3220
53. Larsen MV et al (2005) An integrative approach to CTL epitope prediction: a combined algorithm integrating MHC class I binding, TAP transport efficiency, and proteasomal cleavage predictions. Eur J Immunol 35:2295–2303. https://doi.org/10.1002/eji.200425811
54. Fritsch EF, Hacohen N, Wu CJ (2014) Personal neoantigen cancer vaccines: the momentum builds. Oncoimmunology 3:e29311. https://doi.org/10.4161/onci.29311
55. Wang ET et al (2008) Alternative isoform regulation in human tissue transcriptomes. Nature 456:470–476. https://doi.org/10.1038/nature07509
56. Chen J, Weiss WA (2015) Alternative splicing in cancer: implications for biology and therapy. Oncogene 34:1–14. https://doi.org/10.1038/onc.2013.570
57. Danan-Gotthold M et al (2015) Identification of recurrent regulated alternative splicing

events across human solid tumors. Nucleic Acids Res 43:5130–5144. https://doi.org/10.1093/nar/gkv210
58. Barrett CL et al (2015) Systematic transcriptome analysis reveals tumor-specific isoforms for ovarian cancer diagnosis and therapy. Proc Natl Acad Sci U S A 112:E3050–E3057. https://doi.org/10.1073/pnas.1508057112
59. Eswaran J et al (2013) RNA sequencing of cancer reveals novel splicing alterations. Sci Rep 3:1689. https://doi.org/10.1038/srep01689
60. Zhao Q et al (2013) Tumor-specific isoform switch of the fibroblast growth factor receptor 2 underlies the mesenchymal and malignant phenotypes of clear cell renal cell carcinomas. Clin Cancer Res 19:2460–2472. https://doi.org/10.1158/1078-0432.Ccr-12-3708
61. Oltean S, Bates DO (2014) Hallmarks of alternative splicing in cancer. Oncogene 33:5311–5318. https://doi.org/10.1038/onc.2013.533

Correction to: Considerations in Developing Reporter Gene Bioassays for Biologics

Jamison Grailer, Richard A. Moravec, Zhijie Jey Cheng, Manuela Grassi, Vanessa Ott, Frank Fan, and Mei Cong

Correction to:
Seng-Lai Tan (ed.), *Immuno-Oncology: Cellular and Translational Approaches*, Methods in Pharmacology and Toxicology,
https://doi.org/10.1007/978-1-0716-0171-6_9

The chapter "Considerations in Developing Reporter Gene Bioassays for Biologics" has been made open access under a CC BY 4.0 license and the Copyright Holder is now "The Author(s)". The book has also been updated to reflect this change.

The updated online version of this chapter can be found at
https://doi.org/10.1007/978-1-0716-0171-6_9

Seng-Lai Tan (ed.), *Immuno-Oncology: Cellular and Translational Approaches*, Methods in Pharmacology and Toxicology, https://doi.org/10.1007/978-1-0716-0171-6_14, © Springer Science+Business Media, LLC, part of Springer Nature 2020

INDEX

A

Abcam Kinetic Apoptosis Kit pSIVA™ 106
Acute lymphoid leukemia (AML) 240
Adenosine 75
Adenovirus serotype 5 (Ad5) 227
Alemtuzumab 215
AlexaFluor™ 647 174
Antagonistic antibodies 219
Antibody clone screening 173
Antibody-dependent cellular cytotoxicity (ADCC) 13, 22, 24, 26, 202, 204–207, 221
Antibody drug conjugate (ADC) 13, 15–18
Antigen-based potency assay 74
Antigen challenge models
    adverse drug-related effects 225
    humoral/cellular immune function 224
    ICH S8 guidance 224
    immune cells 226
    immune suppression 224
    inflammatory/autoimmune-type toxicities 224
    intracellular cytokine analysis 225
    in vitro measurement 225
    nonclinical toxicity studies 225
    viral challenge models 225, 227, 228
Antigen presenting cells (APCs) 221, 244
Antigen specific CD8$^+$ T cells 82
Antiidiotype detection antibody 219
Anti-tumoral cell-mediated responses 240
Antitumor immunity 225
Anti-tumor response 11
Artificial neural networks (ANNs) 248
Assay interference 177
Assay optimization
    buffer and media 141, 142
    culturing cells 140, 141
    DoE 142, 144
    hook effect 142
    induction time 141
    plate edge effect 142
    plating cells 141
    standardizing assay reagents 140

B

B-cell maturation antigen (BCMA) 240
BD Cytofix™ 181

BD QuantiBrite™ 179
Bio-layer interferometry (BLI) 223
Biologics
    approaches 131
    assay design
        cell background 133
        genetic reporter 133, 134
        positive control 134
    assay feasibility studies 134, 135
    bioassay 132, 147, 148
    biological activity 132
    bioluminescent reporter gene bioassays 133
    biopharmaceutical drug development 132
    challenges 152, 154, 155
    clone stability testing 138, 139
    developing potency bioassays 132
    development, thaw-and-use cells 143, 146
    drug development 132
    immunotherapy 131
    property 132
    selecting cell clones 136, 137
BioSpa™ 8 Automated Incubator 105
BioSpa method 106
Biotherapeutics 200, 203, 205, 206, 214, 216–218, 222, 223
Bispecific antibody-based cytotoxicity assay 23
Bispecific antibody killing assay 24
Bispecific immunotherapeutics 212
Bispecific modalities 205, 206
Bispecifics 55
Bispecific T cell engagers (BiTE®) 53
BiTE antibody 47
Blood lysing reagent 194
Brightfield imaging 107
Brilliant Violet™ 174

C

Calcein AM 13
Calcein-based imaging 21, 22, 31
Cancer antigens
    cancer progression and metastasis 249
    cancer-specific formation 249
    cancer-specific isoforms 250
    CancerSplice 250
    data-mining tools 246–248

Index

Cancer antigens (*cont.*)
  "driver" neoantigens 244–246
  immunotherapy 250
  KAI's ecosystem 247
  mechanisms 249
  mRNAs 249
  "passenger" neoantigens 243, 244
  T-cell response 248
  tumor-specific peptide 249
  universe of cancer targets 241
  whole transcriptome sequencing 249
Cancer associated fibroblasts (CAF) 119, 122
Cancer immunotherapy (CIT) 13, 36
  binding affinity and potency assays ... 202, 203
  biology and MOA 200
  drugs 199
  functional immunotoxicological assessments 230
  immune-related adverse events 200
  immune system 199
  immunogenicity assessment 221–223
  immunotoxicology assessments 229, 230
  in vitro assays 200, 201
  MHC tetramer assay and ELISpot ... 227, 229
  nonclinical toxicity studies 200
  nontumor tissues 200
  pharmacological responses 200
Cancer-testis antigens (CTAs) 242
CAR T cell-based therapy 27
CAR T cell killing assay 28
CAR T-cells 239
Carboxyfluorescein succinimidyl ester
  (CFSE) 23, 27, 205
CCR5 Specific Internalization RO Assays 195
CD3/CD28 Dynabeads® 90, 93, 95
CD3/CD28 stimulation 95
Celigo Image Cytometer 15, 16, 23, 33
Cell aggregation 123
Cell-based assays 35, 52, 152
Cell death 206
Cell Index (CI) 37, 56
Cell isolation and culture 122
Cell-mediated cytotoxicity (CMC) assay 104
CellTracker Deep Red dye 106
CellTrace Violet (CTV) 23
CellTracker™ Green CMFDA 23
CellTracker™ Violet BMQC 23
Cell viability 92
Checkpoint inhibitors 1, 7, 89
Chimeric antigen receptor (CAR) 13
  E-plate 44
  FACS 43
  generation 42
Chromium^{51} (^{51}Cr) 13
Chronic viral infection *vs.* cancer 74
Clinical trial applications (CTAs) 216

Co-culture-based cytotoxicity assays 13
Coculture medium composition 122
Combination therapies 100
Combined assay technique 104
Complementarity-determining region (CDR) 216
Complement-dependent cytotoxicity
  (CDC) 14, 16, 19, 20, 202, 206, 208
Confocal microscopy 206
Conventional T cell activation assays 100
CryoMed Freezer 146
Cryopreserved NK cells 106
Culture characterization 123
Cytation 5 imaging chamber 105, 106
CytoChex collection tubes 188
Cytokine production 212
Cytokine release syndrome (CRS)
  biologic agent 216
  cardiovascular shock and acute respiratory
    distress syndrome 214
  cynomolgus monkeys 214
  dry-coat method 215
  lymphocytes and monocytes 214
  mild/moderate cytokine release 214
  molecular mechanisms 212
  nonclinical toxicity 214, 215
  PBMC solid phase assay 215
  physiological conditions 212
  proinflammatory cytokines 212
  science-based approach 216
  science-based assessment 214
  solid-phase assay 215
  soluble- and solid-phase assay 214
  systemic inflammatory response 212
Cytotoxicity 28
Cytotoxic T lymphocytes (CTLs) 245

D

Daudi cells 44
Dead Cell Removal Kit 3
Design of experiment (DoE) 142, 144
Dilution experiment design 182
Direct cell-mediated cytotoxicity assay 19
Drug combination 157–159
Drug development 120, 200, 219–223
Drug dose response curves 176
Drug potency 132, 133
Drug response 158, 164
Drug screening 157, 158
Dynabeads® 92, 93

E

EasySep T cell isolation 91
EasySep™ 46, 122
Effector cell activation

cell-engaging antibodies	208
clinical efficacy	208
cytokines	208
cytolytic proteins	208
effector-to-target ratio	211
in vitro assays	209, 210
MOA	209
pharmacologic activity assessments	209
target binding affinity	211–213
target expression	211
T-cell bispecific antibodies	209
TDB	210
and treatment	44–46
Efferocytosis	23, 25
EGFR-CD28 CAR-T cells	43, 44
Electrochemiluminescence immunoassay (ECLIA)	223
Elotuzumab	206
Enzyme-linked immunosorbent assay (ELISA)	27, 202
Enzyme-linked immunospot (ELISpot) assay	227, 229
E-plate	44
Equivalent relative fluorescence (ERF)	183, 189
Erythrocyte removal	124
European Medicines Agency (EMA)	214, 223
Extracellular matrix (ECM)	119

F

Fc effector function assays	
ADCC	204, 205
ADCP	205–207
amino acid modifications	204
blocking/mimicking receptor-ligand interactions	203
CDC assays	206, 208
in vitro cell-based assays	203
in vitro characterization	204
multiple antibody engineering methods	204
types	203
Fetal bovine serum (FBS)	54, 141
Fibrinogen-like Protein 1 (FGL1)	155
First-in-human (FIH) dose	200
Flow cytometric method	206
Flow cytometry	3, 27, 92, 157, 218, 227
Fluorescence-activated cell sorting (FACS)	136
Fluorescent channels	163
Fluorochromes	172, 174
Fluorophore selection	165
Food and Drug Administration (FDA)	223, 240
Free site receptor occupancy assays	
advantages	179, 185
anticoagulated whole blood	181
co-stain antibodies	182
disadvantages	180, 185
FMO	184
isotype control	186
materials	181–180
MESF/ERF	183, 190
neutrophils	185
performance	182
process	185
RO assays	180
temperature	182
therapeutic binding site	180
titration dilution tubes	182
Functional receptor occupancy assays	189–195
advantages	189
anti-drug antibody	192
disadvantages	189
RO testing	189

G

Galvanometric mirrors	14
Gen5™ software	107, 108
Genetically engineered murine models (GEMM)	118
Genotype-tissue expression (GTEx)	246
Granulocyte macrophage colony stimulating factor (GM-CSF)	215
GraphPad Prism 8	92

H

Hardware-Based Auto-Focus (HBAF)	14
HCT116 epithelial colorectal carcinoma cells	105
Highest nonseverely toxic dose (HNSTD)	212
High-pressure liquid chromatography (HPLC)-based methods	223
Human immunodeficiency virus (HIV) pathogenesis	225
Human leukocyte antigen (HLA) targets	24, 53
Human papilloma virus (HPV)	243
Human TruStain FcX™	194
Human tumor immunology	225
Human umbilical vein endothelial cells (HUVEC)	215
Hypoxic conditions	75

I

Image-Based Auto-Focus (IBAF)	14
Image cytometric analysis methods	15
Image pre-processing and stitching	110
ImmTAC molecules	24, 26
Immune checkpoint blockades (ICB)	78, 90
Immune checkpoint drug candidates	
antigen-based potency assay	77

Index

Immune checkpoint drug candidates (*cont.*)
- cancer immunotherapy 75
- cancer-induced suppressive microenvironment 81
- challenges 76, 77
- clinical efficacy 76
- drug development 76
- functional assays 75
- functional screening and assessment 76
- immune-related biomarkers 76
- in vitro functional screening 86
- multiple co-inhibitory receptors 77
- parameters 82
- PBMCs 85, 86
- screening tool 82
- superantigen-based assays 78
- T cell dysfunction 81
- tumor-induced exhaustion agents 84
- tumor-induced T cell dysfunction 85
- utility for functional screening 85

Immune checkpoint inhibitors (ICI) 239
Immune mobilising monoclonal TCR Against Cancer (ImmTAC®) 53
Immuno fluorescent labeling (IF) 161–163
Immunogenicity assessment 221–223
Immunomodulators 78

Immunomodulatory antibodies
- experimental procedures
  - human cancer 3
  - mice 2, 3
- human material 7, 8
- in vitro model 1
- long-term memory T cells 2
- lymphocytes expressing TNFγ 2
- lymphoid cells 1
- mouse SW1 melanoma cells 10
- mouse tumors 4, 5, 7
- non-small cell lung carcinoma 1
- ovarian carcinoma 9
- spleen cells 4
- Th2 related genes 7

Immunomodulatory drug 74
Immuno-oncology 13, 33, 44
Immunotherapeutic agents 1
Immunotherapy 131, 152, 154, 155, 159, 160, 239
Immunotoxicology assessments 229, 230
In silico prediction 246, 250

In vitro cancer cell killing assays
- ACEA Biosciences' xCELLigence system 52
- animal models 51
- assessment 52
- effective anticancer response 52
- effector cells 65, 66
- electrodes 53
- features 65
- fluorescently labeled cells 67
- IncuCyte® 52
- IncuCyte *vs.* xCELLigence methodologies 53, 65
- materials
  - common reagents 54
  - IncuCyte-specific reagents 54
  - multiplexing 54
  - xCELLigence-specific reagents 54
- methods
  - IncuCyte killing assay protocol 57, 59–62
  - optimizing seeding density 56, 58
  - principle of assay design 55, 56
  - seeding density optimization 57
- monoclonal antibodies 53
- nonadherent cells 66
- normal/primary cells 67
- preclinical assessment 52
- RTCA 53
- T cell activation 52
- T cell killing 53

In vitro cytotoxicity assay 211
In vitro functional assay
- cell proliferation 35
- E-plate 39
- materials 37–39
- NK cell-mediated cytotoxicity 39
- xCELLigence RTCA 37
- xCELLigence technology 36

In vitro-generated T_{EX} cells 100
In vitro immunotherapy assays 14, 15, 28, 33
Inducer of T cell suppression (IoTS) 84
Intercellular adhesion molecule 1 (ICAM1) 78
Inter-instrument standardization processes 168
International Consortium of Harmonization (ICH) 223
Intrinsic factors 222
Investigational New Drug (IND) 214–216
iScriptTM Reverse Transcription Supermix 3
Isocitrate dehydrogenase-1 (IDH1) 245

K

Keyhole limpet hemocyanin (KLH) 224
Kinetic cytotoxicity assay 92
Kinetic exclusion assays (KinExA) 202
Kinetic montage images 110
Ki-67 expression 96

L

Lactate dehydrogenase (LDH) 13
LAG-3 .... 81, 82
Leukemia 159, 160, 164

Leukocyte function-associated antigen 1 (LFA-1) 78
Light-emitting diodes (LEDs) 14
Long-term T cell killing.. 32
Luciferase Assay System 133
Luminometer 154
Lymphocytes 6
Lymphocytic choriomeningitis virus
    (LCMV) 74, 78, 90
Lymphoprep™ 122

M

mAb combination 2, 4, 5, 7–9
MABEL-based approach.. 220
Macrophages 119, 120, 205
Major histocompatibility complex
    (MHC)..77, 221
Major histocompatibility complex class II
    (MHC II)..78, 155
Manufacturing elements 223
Master Cell Bank (MCB).. 143
Maximum tolerated dose (MTD) 209
Mechanism of action (MOA) 132, 200
Membrane attack complex (MAC) 206
Mesoscale Discovery (MSD) 70
Microencapsulation procedure 124
Microwell assay chamber 165
Minimum anticipated biological effect
    level (MABEL) approach 168
Mixed lymphocyte reaction (MLR) 75, 77, 90, 92
Modulatory/functional internalization ROA 194
    clone screening 193
    IgG secondary detection antibody......................... 193
Molecules of equivalent soluble fluorescence
    (MESF) 183, 189
Monoclonal antibodies (mAbs).................................. 1–5,
    7, 8, 10, 11, 131, 202
Monocyte-derived dendritic cells (mo-DCs) 92
Monocytes 206
Mononuclear cells (MNC) 161
Mouse immune system 225
Mouse tumor models... 3–5, 7
Mouse tumors 4, 5, 7
Multiplexed endpoint assays
    cytokine analysis 70
    T cell proliferation protocol 68–70
Multitargeting biotherapeutics...................................202

N

Natural killer (NK) cell 39
    anti-cancer activity 103
    BioSpa 105
    cellular analysis steps 108
    cytation™ 5 105
    cytotoxicity imaging....................... 108–107
    fluorescent apoptosis and necrosis probes 104
    HCT116 cells 104, 106
    immunotherapy 103
    matrix infiltration studies....................... 103
    RAFT™ process 106
    and target tumor cells 103
    3D cultured cells 104
    tumoroids 104, 108
Nivolumab 95
NK92 cells 39, 41
2018 Noble Prize in Medicine and Physiology 239
Non-Hodgkin lymphoma (NHL)......................... 39, 240
Non-infiltrated myeloid fraction 126
Non-neutralizing ADA 174
Nonneutralizing antibody 221
Non-small cell lung carcinoma (NSCLC) 1, 122
No-observed adverse effect level (NOAEL)................212
Normalized Cell Index (NCI)...................................... 37

O

Optical biosensor 202
Ovarian carcinoma 8

P

PBMC-based culture system 82
PBMC cell suspension 46
PD-1 ....77, 78, 81, 82, 86
Pembrolizumab 92, 98
Peptide-HLA complex (pHLA) 61
Peripheral blood-derived monocytes isolation 124
Peripheral blood mononuclear cells
    (PBMCs) 3, 46, 54, 91, 122, 215
Phagocytosis 23
Pharmacodynamic (PD) profile 219
Phenotypic assays 35
Phosphate buffered saline (PBS).................................122
Phosphatidyl serine (PS) 108
Potency Score 82
Proinflammatory cytokines 78
Propidium iodide (PI) probe 108

Q

Qiagen RNeasy Mini Kit 3
Qualifying potency bioassays
    assay qualification 149
    parameters 150
    qualification report.................................. 152
    reporter gene bioassay 149
    system suitability acceptance criteria
        agreement 152
        analytical procedure 152
        definition 150

Index

intermediate precision expresses 152
linearity 153
nominal drug concentration 152
optimization/prequalification experiments 151
parallelism 151
parameters 151
qualification parameters 151
range, analytical procedure 153
reporter gene bioassay 150, 152
specificity 151
Quantitative analysis 108

R

Radioimmunoassay (RIA) 223
RAFT™ 3D cell culture system 111
RAFT™ 96-well Small Kit 105
RAFT™ hydrogel matrix 107
Real-time cell analyzer (RTCA) system 36, 53, 92
    assays and analyses 41
    E-plate 41
    software 41
    solid tumors 42
    tumor cell lines 42
    xCELLigence 36
Recall antigen assay (RAA) 75
    chronic infection model 80
    co-stimulatory/co-inhibitory molecules 80
    functional screening 80
    function-based screening 79
    optimizations and validation 78
    positive and isotype antibodies 80
    qualification and validation 79
    tetramers 80
Receptor occupancy (RO) assays 218–221
    ADA 168
    amine reactive conjugations 173
    antibody therapeutics 170
    anti-CD38 RO assays 173
    anti-CD38 therapy competition testing 172
    anti-coagulant 177
    anti-idiotype antibodies 172
    anti-PD-1 and anti-CD38 therapy 172
    biotinylation 174
    CD3 and CD4 170
    design and implementation 168
    flow cytometry 167
    fluorochrome 179
    longitudinal 169
    MESF 176
    modulatory/functional assays 172
    PBMC cell suspension 171
    PBMCs 170
    PBMC stimulation 170
    phase I clinical trials 168
    phase II and phase III clinical trials 168
    quasi-quantitative 178
    redirecting T cells 169
    saturation curve 169
    semi-quantitative/quasi-quantitative methods 178
    stability 177
    target engagement assays 167
    technical and logistical challenges 168
    temperatures 177
    testing 168
    therapeutic 172
    traditional 169
Receptor targets 187
Relative luminescent unit (RLU) signal 137
Relative potency (RP) 149
Reporter bioassay 143, 147, 150–152
Reporter gene bioluminescence assays 207
Representative data 196
Restimulation model 98

S

Safety assessment 200, 214, 217
Signal amplification methods 170
Simian immunodeficiency virus (SIV)
    components 225
Single cell imaging
    cell line selection 164
    drug conditions 158
    ex vivo screening platforms 158
    flow cytometry 157
    fluorescence microscopy 159
    fluorescence quantitation 159
    fluorophore selection 165
    genetic-mutational information 159
    image acquisition and quality assessment 165
    imaging assay approach 158
    materials
        cell culture 160
        cell preparation 159
        IF 161
        image acquisition 161
        image analysis 161
    methods
        cell culture 162
        cell preparation 161
        IF 162, 163
        image acquisition 163
        image analysis 164
    microwell assay chamber 165
    MOLM 14 cell 164
    myeloid and T-cells 159
    negative detection 158
    primary patient cells 157

reagent optimization 165
technical capabilities 158
Single RNAseq 100
SKOV-3/HeLa cells 92
Spheroid concentration 125
Spheroid's diameter 125
Splenocytes 4
Staphylococcal enterotoxin B (SEB) 75, 78
Stirred-tank vessel 126
Superantigen-based assays 78
Surface plasmon resonance (SPR) 202, 223

T

T cell
    activation and proliferation measurement 27, 90
    dysfunction assay 82
    migration 27
    therapy 13
T cell exhaustion 74, 81, 83
    activation 89
    antibodies 91
    ATCC 91
    cultured T cells 92
    cytotoxicity 96
    dysfunction 89
    dysfunctional nature 96
    PD-1/PD-L1 pathway 95
    RTCA 92
    secondary assays 90
    T_{EX} cells 93
T cell killing assay
    IncuCyte platform 61, 63
    xCELLigence platform 63, 64
T cell receptor (TCR) 53, 77, 78, 93
T-cell-dependent antibody response
    (TDAR) 224, 225
T-cell-dependent bispecific antibody
    (TDB) 210
TDAR-SIV model 227, 228
Tebentafusp 53
Tetramers 227
T_{EX} cells 91
Th1 response 5, 10
Th2 response 7
Thaw-and-use (T&U) cells 143, 146
3D cell/collagen hydrogel 104
3D cytotoxicity fold induction calculation 111
3D NK CMC analysis 110
3D tumoroid formation conformational
    imaging 107
3D-3-culture 123
    alginate 119
    animal models and clinical trials 120
    cell morphology and polarity 118

characterization 126
coculture model 120
cytokines 120
factors 118
human-relevant immune microenvironment 118
immunogenicity 119
immunotherapies 120
in vitro models 118
M1/M2 paradigm 120
M2-like immunosuppressive phenotype 119
macrophage phenotype 120
pro-inflammatory phenotype 120
reagents and equipment 121
stirred-tank systems 119
TME 119, 120
treatment improvement 118
tumor and immune cells 119
TIM-3 .... 77, 81, 82
Tissue cross-reactivity (TCR) studies
    assessment 217
    BiAbs/TDBs 216
    CIT molecules 217, 218
    complex tissue microenvironment 217
    efficacy studies 217
    immunohistochemistry techniques 216
    limitations 217
    outcomes 216
    positive control tissue 216
Tissue-infiltrating lymphocytes (TILs) 90
Titration dilution tubes 189
Titration experiments 191, 192
T lymphocytes (T-cells) 162
TME suppressive mediators 75
TruStain FcX™ 181, 188
Tumor antigens 53
Tumor-associated antigens (TAAs)
    cancer immunotherapy 241
    categories 240, 241
    cellular and molecular techniques 242
    development of genetic methods 242
    immune system–tumor cell interactions 241
    lymphocytes 242
    oncofetal antigens 242
    overexpressed antigens 242
    proteins/glycoproteins 240
    structural protein 242
    subcellular compartment 240
    T lymphocytes 242
    TSAs 243
    virus-derived proteins 243
Tumor associated macrophages (TAM) 119
Tumor cell line 122
Tumor-destructive Th1 response 10
Tumor-induced exhaustion agents 84

Index

Tumor microenvironment (TME) 23, 120
    characteristic 120
    components 120
Tumor-specific antigens (TSAs) 243
Two-color fluorescence 61
2D cytotoxicity fold induction calculation 114
2D NK cell-mediated cytotoxicity imaging 110
2D NK CMC analysis 113

V

V-domain immunoglobulin suppressor of T
    cell activation (VISTA) 155
Viral challenge models 225, 227, 228

W

Whole exome sequencing (WES) 243